An Engineering Data Book

An Engineering Data Book

Third edition

Edited by
JR Calvert and RA Farrar

First published 2008 by
PALGRAVE MACMILLAN

Palgrave Macmillan in the UK is an imprint of Macmillan Publishers Limited,
registered in England, company number 785998, of Houndmills, Basingstoke,
Hampshire RG21 6XS.

Palgrave Macmillan in the US is a division of St. Martin's Press LLC,
175 Fifth Avenue, New York, NY 10010.

Palgrave Macmillan is the global academic imprint of the above companies
and has companies and representatives throughout the world.

Palgrave® and Macmillan® are registered trademarks in the United States,
the United Kingdom, Europe and other countries.

ISBN-13: 978–0–230–22033–1   paperback

This book is printed on paper suitable for recycling and made from fully
managed and sustained forest sources. Logging, pulping and manufacturing
processes are expected to conform to the environmental regulations of the
country of origin.

A catalogue record for this book is available from the British Library.

A catalog record for this book is available from the Library of Congress.

Contents

Preface

Learning formulae is the bugbear of the student, and many avoidable errors, both in education and professional practice, have been made through the incorrect recall of formulae or numbers. This booklet aims to provide a ready reference for commonly required formulae and data, for use in coursework and examinations (where permitted) and in professional practice. It is not a textbook – the user is expected to know when to use a particular formula, and just as important, when not to use it.

The units used are SI, or multiples, with conversion factors from other systems provided. The symbols are generally those in common use in particular specialities, except where this would cause confusion in context in a particular section of the book. A complete list of symbols, with meaning and units, is included.

This edition is an extended and revised version of that edited by AJ Munday and RA Farrar, published in 1979 in collaboration with The Macmillan Press Ltd, itself based on an internal publication of the Department of Mechanical Engineering, University of Southampton.

Data have been collected from a large variety of sources, and the editors are grateful to many colleagues, past and present, for their contributions and suggestions.

The editors have made every effort to ensure the accuracy of this data, but cannot guarantee it. No responsibility can be taken for the consequences of any errors which may remain. Any person finding an error is asked to inform the publisher.

Preface to the 2008 edition

The main change in this new edition is the inclusion of a new section "Earth and the Environment". There are also numerous smaller additions throughout the work, and some entries have been changed slightly to improve clarity or match current practice better. Nothing has been removed. Once more, we owe a debt of gratitude to the many colleagues who have made suggestions and offered constructive criticism on the presentation of the data.

JRC
Southampton, March 2008

1. Symbols and Units
1.1 *Symbols*

The Greek Alphabet

A	α	alpha	H	η	eta	N	ν	nu	T	τ	tau
B	β	beta	Θ	θ	theta	Ξ	ξ	xi	Y,Υ	υ	upsilon
Γ	γ	gamma	I	ι	iota	O	o	omicron	Φ	ϕ	phi
Δ	δ	delta	K	κ	kappa	Π	π	pi	X	χ	chi
E	ε,ϵ	epsilon	Λ	λ	lambda	P	ρ	rho	Ψ	ψ	psi
Z	ζ	zeta	M	μ	mu	Σ	σ	sigma	Ω	ω	omega

Mathematical Symbols

L[ ]	Laplace transform	$\underline{a}$	Vector
$\underline{\Delta}$	Defined as	$\hat{\underline{a}}$	Unit vector
Σ	Repeated summation	$\perp$	'at right angles to'
Π	Repeated multiplication	$\bullet$	Scalar (dot) product
∂	Partial differentiation	$\times, \wedge$	Vector (cross) product
lal	Modulus	$\cup$	Union (set), 'or'
∇	Laplacian operator: Del, Nabla	$\cap$	Intersection (set), 'and'
Re( )	Real part of complex number	$\vert$	Given that (probability)
Im( )	Imaginary part of complex number		

Decimal Prefixes

Symbol	Prefix	Multiplier	Symbol	Prefix	Multiplier
T	tera	10^{12}	c	centi	10^{-2}
G	giga	10^{9}	m	milli	10^{-3}
M	mega	10^{6}	μ	micro	10^{-6}
k	kilo	10^{3}	n	nano	10^{-9}
h	hecto	10^{2}	p	pico	10^{-12}
da	deca	10	f	femto	10^{-15}
d	deci	10^{-1}	a	atto	10^{-18}

1.2 *SI Units*

Basic Units

Quantity	Unit Name	Unit Symbol
Length	metre	m
Mass	kilogramme	kg
Time	second	s
Amount of Substance	mole	mol
Electric Current	ampère	A
Thermodynamic Temperature	kelvin	K
Luminous Intensity	candela	cd

Supplementary and Derived Units

Quantity	Unit Name	Unit Symbol	Equivalent Units	Dimensions MLT
Plane Angle	radian	rad	-	-
Solid Angle	steradian	sr	-	-
Force	newton	N	$kg\ m/s^2$	MLT^{-2}
Work, Energy, Heat	joule	J	$N\ m$	ML^2T^{-2}
Power	watt	W	$J/s, Nm/s$	ML^2T^{-3}
Frequency	hertz	Hz	s^{-1}	T^{-1}
Dynamic Viscosity	-	Ns/m^2	$kg/m\ s$	$ML^{-1}T^{-1}$
Kinematic Viscosity	-	m^2/s	-	L^2T^{-1}
Pressure	pascal	Pa	N/m^2	$ML^{-1}T^{-2}$

Electrical and Magnetic

Quantity	Unit Name	Unit Symbol	Equivalent Units
Potential	volt	V	$W/A, Nm/As$
Resistance	ohm	Ω	V/A
Charge	coulomb	C	$A\ s$
Current	ampère	A	C/s
Capacitance	farad	F	$A\ s/V$
Electric Field Strength	-	V/m	-
Electric Flux Density	-	C/m^2	-
Magnetic Flux	weber	Wb	$V\ s, Nm/A$
Inductance	henry	H	$V\ s/A, Nm/A^2$
Magnetic Field Strength	-	A/m	
Magnetic Flux Density	tesla	T	$Wb/m^2, N/(Am)$

Light

Quantity	Unit Name	Unit Symbol	Equivalent Units
Luminous Flux	lumen	lm	$cd\ sr$
Luminance	nit	cd/m^2	-
Illumination	lux	lx	lm/m^2

1.3 *Conversion Factors for some other units into SI Units*

Length

1 Å (ångstrom)	=	10^{-10} m			1 μ (micron)	=	1 μm
1 in (inch)	=	25.4 mm (exactly)			1 μin	=	0.0254 μm
1 thou	=	1 mil	=	0.001 in	=	25.4 μm
1 ft (foot)	=	0.3048 m			1 yd (yard)	=	0.914 m
1 mi (mile)	=	5280 ft	=	1.609 km			
1 nm (nautical mile)	=	6076 ft	=	1.852 km (exactly)	≈	1 min of latitude along a meridian
1 km	=	0.622 mi	=	0.540 nm			

Area

1 acre	=	0.4047 ha (hectare)	=	4 047 m^2
1 ha	=	10 000 m^2		

Volume

1 in^3	=	16.39 cm^3			1 ft^3	=	0.02832 m^3
1 l (litre)	=	1 dm^3	=	10^{-3} m^3			
1 cc	=	1 cm^3	=	1 ml	=	10^{-6} m^3
1 gal (imperial gallon)	=	0.1605 ft^3	=	4.546 l	=	4546 cm^3
1 US gal	=	0.1337 ft^3	=	3.785 l	=	3785 cm^3

Mass

1 tonne (metric ton)	=	1 Mg	=	1000 kg
1 lb (pound)	=	0.45359237 kg (exactly)	1 kg	=	2.205 lb
1 slug	=	32.17 lb	=	14.59 kg
1 ton	=	2240 lb	=	1016 kg
1 short ton	=	2000 lb	=	907.2 kg

Density

1 lb/in^3	=	27.68 g/cm^3	=	27 680 kg/m^3
1 lb/ft^3	=	16.02 kg/m^3	1 g/cm^3	=	1000 kg/m^3
1 $slug/ft^3$	=	515.4 kg/m^3		

Velocity, angular velocity

1 mile/h	=	1.467 ft/s	=	1.609 km/h	=	0.447 m/s
1 knot	=	1 nm/h	=	1.853 km/h	=	0.514 m/s
	=	1.689 ft/s				
1 rev/s	=	6.283 rad/s		1000 rev/min	=	104.7 rad/s

Flow rate

1 ft^3/s	=	1 cusec	=	28.32 l/s	=	0.02832 m^3/s
1 cufm (cubic foot per minute)	=	28.32 l/min	=	1.699 m^3/h		
1 gal/min	=	0.07577 l/s	=	7.577×10^{-5} m^3/s		

Force, Weight

1 dyne	=	10^{-5} N		1 tonf (ton force)	=	9964 N
1 pdl (poundal)	=	0.1383 N				
1 lbf (pound force)	=	32.17 pdl	=	4.448 N		
1 kp (kilopond)	=	1 kgf	=	2.205 lbf	=	9.807 N

Torque

1 lbf in	=	11.3 N cm	=	0.113 Nm
1 lbf ft	=	1.356 Nm		
1 tonf ft	=	3037 Nm		

Stiffness

1 lbf/μ in	=	175 N/μm		1 lbf/in	=	17.5 N/m

Energy, Work, Heat

1 eV (electron volt)	=	1.602×10^{-19} J		1 ft lbf	=	1.356 J
1 kW h	=	3.6 MJ		1 hp h	=	2.685 MJ
1 cal (calorie)	=	4.187 J				
1 Btu (British thermal unit)	=	778.2 ft lbf	=	252 cal	=	1055 J
1 Chu (Centigrade heat unit)	=	1.8 Btu	=	1899 J		

Power, Heat Flow

1 ft lbf/s	=	1.356 W		1 Btu/h	=	0.293 W
1 hp (horsepower)	=	550 ft lbf/s	=	33 000 ft lbf/min	=	0.7457 kW
1 ch (metric horsepower)	=	1 PS	=	0.7355 kW		

Pressure, Vacuum, Stress

1 lbf/in^2	=	0.07031 kgf/cm^2	=	6895 N/m^2		
1 lbf/ft^2	=	47.88 N/m^2				
1 tonf/in^2	=	157.5 kgf/cm^2	=	15.44 MN/m^2		
1 kgf/cm^2	=	0.09807 MN/m^2	=	0.9807 bar		
1 dyne/cm^2	=	0.1 Pa	=	0.1 N/m^2		
1 kgf/mm^2	=	9.807 MN/m^2	=	98.07 bar		
1 bar	=	14.50 lbf/in^2	=	0.1 MPa	=	10^5 N/m^2
1 atm (international atmosphere)	=	14.70 lbf/in^2	=	760 torr (exactly)	=	101.325 kPa
			=	1.01325 bar		(exactly)
1 torr	=	133.3 Pa	=	133.3 N/m^2		

Stress Intensity

$1 \mathrm{ksi}\sqrt{\mathrm{in}}$ (kilo-pound per square inch)	=	1.10 MNm$^{-3/2}$

Head

1 ft H$_2$O	$\approx$	62.43 lbf/ft^2	=	2989 Pa
1 cm H$_2$O	$\approx$	98.1 Pa		
1 in Hg	=	13.6 in H$_2$O	$\approx$	3388 Pa
1 mm Hg	$\approx$	1 torr	=	133.3 N/m^2
1 atm	=	1.01325 bar	$\approx$	10.34 m water
			$\approx$	760 mm Hg

Dynamic Viscosity

1 P (poise)	=	1 g/(cm s)	=	0.1 Ns/m^2
1 cP	=	10^{-3} kg/(m s)	=	1 mN s/m^2
1 kgf s/m^2	=	9.807 N s/m^2		
1 lb/ft h	=	0.4132 mN s/m^2		
1 slug/ft s	=	1 lbf s/ft^2	=	47.88 N s/m^2
1 Reyn	=	1 lbf s/in^2	=	6895 N s/m^2

Kinematic Viscosity

1 ft^2/s	=	0.09290 m^2/s		
1 in^2/s	=	645.2 mm^2/s		
1 St (stokes)	=	1 cm^2/s	=	10^{-4} m^2/s
1 cSt	=	1 mm^2/s	=	10^{-6} m^2/s

Temperature

$$T\,°C = \frac{5}{9}(T\,°F - 32) \qquad T\,°F = \frac{9}{5}(T\,°C) + 32$$

$$(\Delta T\,)°C = (\Delta T)K = \frac{5}{9}(\Delta T\,)°F = \frac{5}{9}(\Delta T\,)°R$$

$$T\,K = \frac{5}{9}(T\,°F + 459.67) = \frac{5}{9}T\,°R = T\,°C + 273.15$$

Specific Energy (internal energy, enthalpy, latent heat, calorific value)

1 Btu/lb	=	2.326 kJ/kg
1 ft lbf/lb	=	2.989 J/kg

Specific Entropy, Specific Heat

1 Btu/(lb °R)	=	4.187 kJ/(kg K)

Thermal Conductivity

1 Btu/(ft h °R)	=	1.731 J/(m s °C)	=	1.731 W/(m K)
1 cal/(cm s K)	=	418.7 J/(m s °C)	=	418.7 W/(m K)

Light

1 lm/ft^2	=	1 foot-candle	=	10.763 lx
1 cd/ft^2	=	10.764 cd/m^2		
1 foot-lambert	=	3.426 cd/m^2		

Concentration

1%	=	10‰	=	10 ppt*	=	1 percent
1‰	=	1 ppt*	=	0.1%	=	1 permil
	=	1000 ppm				
1 ppm	=	1000 ppb				

*Caution: ppt here stands for 'parts per thousand'. It may also be used for 'parts per trillion'.

Molar Solution: a solution containing 1 mol of solute in 1 litre of solvent

Normal Solution: a solution containing 1 mol of the active ion (normally H^+ or OH^- or e^-) in 1 litre of solvent.

2.   Physical Constants

Avogadro's number	N	=	6.022×10^{23}/mol
Bohr magneton	β	=	9.274×10^{-24} A m^2
Boltzmann's constant	k	=	1.381×10^{-23} J/K
Stefan-Boltzmann constant	σ	=	5.67×10^{-8} W/(m^{2}K^4)
electron charge	e	=	1.602×10^{-19} C
electronic rest mass	m_e	=	9.109×10^{-31} kg
electronic charge to mass ratio	e/m_e	=	1.759×10^{11} C/kg
Faraday constant	F	=	9.65×10^4 C/mol
permeability of free space	μ_0	=	$4\pi \times 10^{-7}$ H/m
permittivity of free space	ε_0	=	8.854×10^{-12} F/m
characteristic impedance of free space	Z_0	=	$(\mu_0 / \varepsilon_0)^{\frac{1}{2}} = 120\pi\,\Omega$
Planck's constant	h	=	6.626×10^{-34} J s
proton mass	m_p	=	1.672×10^{-27} kg
proton to electron mass ratio	m_p/m_e	=	1836.1
standard gravitational intensity	g	=	9.80665 N/kg
(acceleration)		=	9.80665 m/s^2
unified atomic mass constant	u	=	1.661×10^{-27} kg
universal constant of gravitation	G	=	6.67×10^{-11} N m^2/kg^2
universal gas constant	R_0	=	8.314 J/(mol K)
velocity of light in vacuo	c	=	2.9979×10^8 m/s
		=	$1/\sqrt{\mu_0 \varepsilon_0}$
volume of ideal gas at 1 atm, 0°C (Standard Temp. and Pressure – STP)		=	0.02241 m^3/mol

3. Analysis
3.1 *Vector Algebra*

$$\hat{\underline{a}} = \frac{\underline{a}}{|\underline{a}|}$$

$$\underline{a} = a_1\underline{i} + a_2\underline{j} + a_3\underline{k} \equiv (a_1, a_2, a_3)$$

$$a \equiv |\underline{a}| = \sqrt{a_1^2 + a_2^2 + a_3^2}$$

$$\underline{a} + \underline{b} = (a_1 + b_1, a_2 + b_2, a_3 + b_3)$$

Scalar (dot) Product:

$$\underline{a} \bullet \underline{b} = (a_1 b_1 + a_2 b_2 + a_3 b_3) = ab\cos\theta = \underline{b} \bullet \underline{a}$$

$$\underline{a} \bullet \underline{a} = a^2$$

Vector (cross) Product:

$$\underline{a} \times \underline{b} = \begin{vmatrix} i & j & k \\ a_1 & a_2 & a_3 \\ b_1 & b_2 & b_3 \end{vmatrix} = ab\sin\theta\,\hat{n}, \quad \text{where } \hat{\underline{n}} \perp \underline{a}, \ \hat{\underline{n}} \perp \underline{b}$$

$$= -\underline{b} \times \underline{a}$$

$$\underline{a} \times \underline{a} = 0$$

Moment: $\qquad\qquad \underline{M} = \underline{r} \times \underline{F}$

Velocity: $\qquad\qquad \underline{V} = \underline{\omega} \times \underline{r}$

Triple Scalar Product:

$$[\underline{a}\,\underline{b}\,\underline{c}] = \underline{a} \bullet \underline{b} \times \underline{c} = \underline{a} \times \underline{b} \bullet \underline{c} = \begin{vmatrix} a_1 & a_2 & a_3 \\ b_1 & b_2 & b_3 \\ c_1 & c_2 & c_3 \end{vmatrix}$$

Triple Vector Product:

$$\underline{a} \times (\underline{b} \times \underline{c}) = (\underline{a} \bullet \underline{c})\underline{b} - (\underline{a} \bullet \underline{b})\underline{c}$$

$$(\underline{a} \times \underline{b}) \times \underline{c} = (\underline{a} \bullet \underline{c})\underline{b} - (\underline{b} \bullet \underline{c})\underline{a}$$

Differentiation of Vectors:

$$\frac{d}{dt}(\underline{a} + \underline{b}) = \frac{d\underline{a}}{dt} + \frac{d\underline{b}}{dt} \qquad\qquad \frac{d}{dt}(f\underline{a}) = \frac{df}{dt}\underline{a} + f\frac{d\underline{a}}{dt}$$

$$\frac{d}{dt}(\underline{a} \bullet \underline{b}) = \underline{a} \bullet \frac{d\underline{b}}{dt} + \frac{d\underline{a}}{dt} \bullet \underline{b} \qquad\qquad \frac{d}{dt}(\underline{a} \times \underline{b}) = \underline{a} \times \frac{d\underline{b}}{dt} + \frac{d\underline{a}}{dt} \times \underline{b}$$

$$\frac{d}{dt}(\underline{a} \bullet \underline{b} \times \underline{c}) = \frac{d\underline{a}}{dt} \bullet \underline{b} \times \underline{c} + \underline{a} \bullet \frac{d\underline{b}}{dt} \times \underline{c} + \underline{a} \bullet \underline{b} \times \frac{d\underline{c}}{dt}$$

Equation of a Straight Line: **Equation of a Plane:**

$$\underline{r} = \underline{a} + \lambda \underline{b}$$ $$(\underline{r} - \underline{a}) \bullet \underline{c} = 0 \text{ or } \underline{r} \bullet \underline{c} = \beta$$

Gradient of a scalar:

$$\text{grad } V = \nabla V = \underline{i} \frac{\partial V}{\partial x} + \underline{j} \frac{\partial V}{\partial y} + \underline{k} \frac{\partial V}{\partial z} \quad \text{(Cartesian)}$$

$$= \underline{u}_r \frac{\partial V}{\partial r} + \underline{u}_\phi \frac{1}{r} \frac{\partial V}{\partial \phi} + \underline{u}_z \frac{\partial V}{\partial z} \quad \text{(Cylindrical)}$$

where:

$$\underline{u}_r = \underline{i} \cos \phi + \underline{j} \sin \phi$$
$$\underline{u}_\phi = -\underline{i} \sin \phi + \underline{j} \cos \phi$$
$$\underline{u}_z = \underline{k}$$

$$= \underline{u}_r \frac{\partial V}{\partial r} + \frac{\underline{u}_\theta}{r} \frac{\partial V}{\partial \theta} + \frac{\underline{u}_\phi}{r \sin \theta} \frac{\partial V}{\partial \phi} \quad \text{(Spherical)}$$

where:

$$\underline{u}_r = \underline{i} \cos \phi \sin \theta + \underline{j} \sin \phi \sin \theta + \underline{k} \cos \theta$$
$$\underline{u}_\theta = \underline{i} \cos \phi \cos \theta + \underline{j} \sin \phi \cos \theta - \underline{k} \sin \theta$$
$$\underline{u}_\phi = -\underline{i} \sin \phi + \underline{j} \cos \phi$$

Divergence of a vector:

$$\text{div } F = \nabla \bullet \underline{F} = \frac{\partial F_x}{\partial x} + \frac{\partial F_y}{\partial y} + \frac{\partial F_z}{\partial z} \quad \text{(Cartesian)}$$

$$= \frac{1}{r} \frac{\partial}{\partial r}(rF_r) + \frac{1}{r} \frac{\partial F_\phi}{\partial \phi} + \frac{\partial F_z}{\partial z} \quad \text{(Cylindrical)}$$

$$= \frac{1}{r^2} \frac{\partial}{\partial r}(r^2 F_r) + \frac{1}{r \sin \theta} \frac{\partial}{\partial \theta}(F_\theta \sin \theta) + \frac{1}{r \sin \theta} \frac{\partial F_\phi}{\partial \phi} \quad \text{(Spherical)}$$

3-2

Curl of a vector:

$$\text{curl}\,\underline{F} = \text{rot}\,\underline{F} = \nabla \times \underline{F} = \underline{i}\left(\frac{\partial F_z}{\partial y} - \frac{\partial F_y}{\partial z}\right) + \underline{j}\left(\frac{\partial F_x}{\partial z} - \frac{\partial F_z}{\partial x}\right) + \underline{k}\left(\frac{\partial F_y}{\partial x} - \frac{\partial F_x}{\partial y}\right)$$

$$= \begin{vmatrix} \underline{i} & \underline{j} & \underline{k} \\ \dfrac{\partial}{\partial x} & \dfrac{\partial}{\partial y} & \dfrac{\partial}{\partial z} \\ F_x & F_y & F_z \end{vmatrix} \qquad \text{(Cartesian)}$$

$$= \frac{1}{r}\begin{vmatrix} \underline{u}_r & r\underline{u}_\phi & \underline{u}_z \\ \dfrac{\partial}{\partial r} & \dfrac{\partial}{\partial \phi} & \dfrac{\partial}{\partial z} \\ F_r & rF_\phi & F_z \end{vmatrix} \qquad \text{(Cylindrical)}$$

$$= \frac{1}{r^2\sin\theta}\begin{vmatrix} \underline{u}_r & r\underline{u}_\theta & r\sin\theta\,\underline{u}_\phi \\ \dfrac{\partial}{\partial r} & \dfrac{\partial}{\partial \theta} & \dfrac{\partial}{\partial \phi} \\ F_r & rF_\theta & r\sin\theta\,F_\phi \end{vmatrix} \qquad \text{(Spherical)}$$

Laplace's Equation:

$$\nabla \bullet \nabla V \;=\; \nabla^2 V = \frac{\partial^2 V}{\partial x^2} + \frac{\partial^2 V}{\partial y^2} + \frac{\partial^2 V}{\partial z^2} \qquad \text{(Cartesian)}$$

$$= \frac{1}{r}\frac{\partial}{\partial r}\left(r\frac{\partial V}{\partial r}\right) + \frac{1}{r^2}\frac{\partial^2 V}{\partial \phi^2} + \frac{\partial^2 V}{\partial z^2} \qquad \text{(Cylindrical)}$$

$$= \frac{1}{r^2}\frac{\partial}{\partial r}\left(r^2\frac{\partial V}{\partial r}\right) + \frac{1}{r^2\sin\theta}\frac{\partial}{\partial \theta}\left(\sin\theta\frac{\partial V}{\partial \theta}\right) + \frac{1}{r^2\sin^2\theta}\frac{\partial^2 V}{\partial \phi^2}$$

$$\text{(Spherical)}$$

Space Curves:

$$\underline{v} = \hat{\underline{u}}\frac{ds}{dt} \qquad\qquad s = \text{arc length, } \hat{\underline{u}} = \text{unit tangent}$$

$$\underline{a} = \frac{v^2}{\rho}\hat{\underline{n}} + \frac{dv}{dt}\hat{\underline{u}} \qquad\qquad \hat{\underline{n}} = \text{unit 'inward' normal}$$

$$\frac{d\hat{\underline{u}}}{ds} = \frac{1}{\rho}\hat{\underline{n}} \qquad\qquad \rho = \text{radius of curvature}$$

$$\hat{\underline{b}} = \hat{\underline{u}} \times \hat{\underline{n}} \qquad\qquad \underline{b} = \text{binormal vector}$$

$$\frac{d\underline{b}}{ds} = \frac{1}{\tau}\hat{\underline{n}}, \qquad \frac{d\hat{\underline{n}}}{ds} = \frac{1}{\tau}\underline{b} - \frac{1}{\rho}\hat{\underline{u}}, \qquad \frac{1}{\tau} = \text{torsion}$$

Identities:

$$\nabla \bullet \phi \underline{u} = \phi \nabla \bullet \underline{u} + \underline{u} \bullet \nabla \phi$$

$$\nabla \times \phi \underline{u} = \phi \nabla \times \underline{u} + \nabla \phi \times \underline{u}$$

$$\nabla \bullet \underline{u} \times \underline{v} = \underline{v} \bullet \nabla \times \underline{u} - \underline{u} \bullet \nabla \times \underline{v}$$

3.2 *Series:*

Standard Series:

$$(b+x)^{\alpha} = b^{\alpha} + b^{\alpha-1}\alpha x + b^{\alpha-2}\frac{\alpha(\alpha-1)}{2!}x^2 + \dots \text{ for } |x| < b$$

$$(1+x)^{\alpha} = 1 + \alpha x + \frac{\alpha(\alpha-1)}{2!}x^2 + \frac{\alpha(\alpha-1)(\alpha-2)}{3!}x^3 + \dots$$
$$\text{for arbitrary } \alpha, \, |x| < 1$$

$$e^x = 1 + x + \frac{x^2}{2!} + \dots + \frac{x^n}{n!} + \dots \text{ for all } x; \qquad e^x = \lim_{n \to \infty}\left(1+\frac{x}{n}\right)^n$$

$$\cos x = 1 - \frac{x^2}{2!} + \frac{x^4}{4!} - \dots + \frac{(-1)^n}{(2n)!}x^{2n} + \dots \text{ for all } x$$

$$\sin x = x - \frac{x^3}{3!} + \frac{x^5}{5!} - \dots + \frac{(-1)^n}{(2n+1)!}x^{2n+1} + \dots \text{ for all } x$$

$$\tan x = x + \frac{x^3}{3} + \frac{2x^5}{15} + \frac{17x^7}{315} + \dots \text{ for } |x| < \frac{\pi}{2}$$

$$\ln(1+x) = x - \frac{x^2}{2} + \frac{x^3}{3} - \dots + \frac{(-1)^n}{(n+1)}x^{n+1} + \dots \text{ for } -1 < x \leq 1$$

Taylor's Series

$$f(a+h) = f(a) + hf'(a) + \frac{h^2}{2}f''(a) + \dots + \frac{h^{n-1}}{(n-1)!}f^{(n-1)}(a) + \frac{h^n}{n!}f^{(n)}(c)$$
$$\text{where } a < c < a+h$$

Maclaurin's Theorem:

$$f(x) = f(0) + xf'(0) + \frac{x^2}{2!}f''(0) + \dots + \frac{x^{n-1}}{(n-1)!}f^{(n-1)}(0) + \frac{x^n}{n!}f^{(n)}(\theta x)$$
$$\text{where } 0 < \theta < 1$$

Stirling's formula for *n*!

For large n: $\quad n! \approx \sqrt{(2\pi)}\, n^{n+\frac{1}{2}}e^{-n}$

or

$$\log_{10} n! \approx 0.39909 + (n + \tfrac{1}{2})\log_{10} n - 0.43429n$$
$$\ln n! \approx 0.91894 + (n + \tfrac{1}{2})\ln n - n$$

Fourier Series:

General Formulae:

If $f(x)$ is periodic, <u>of period 2L</u>, $f(x + 2L) = f(x)$

$$f(x) = \tfrac{1}{2}a_0 + \sum_{n=1}^{\infty} a_n \cos \frac{n\pi x}{L} + \sum_{n=1}^{\infty} b_n \sin \frac{n\pi x}{L}$$

where:

$$a_n = \frac{1}{L} \int_{-L}^{L} f(x) \cos \frac{n\pi x}{L} \, dx \qquad\qquad n = 0, 1, 2, \ldots$$

$$b_n = \frac{1}{L} \int_{-L}^{L} f(x) \sin \frac{n\pi x}{L} \, dx \qquad\qquad n = 1, 2, 3, \ldots$$

If $f(x)$ is an <u>even</u> function of x (i.e. $f(-x) = f(x)$), then

$$a_n = \frac{2}{L} \int_{0}^{L} f(x) \cos \frac{n\pi x}{L} \, dx \qquad\qquad n = 0, 1, 2, \ldots$$

$$b_n = 0 \qquad\qquad n = 1, 2, 3, \ldots$$

If $f(x)$ is an <u>odd</u> function of x (i.e. $f(-x) = -f(x)$), then

$$a_n = 0 \qquad\qquad n = 0, 1, 2, \ldots$$

$$b_n = \frac{2}{L} \int_{0}^{L} f(x) \sin \frac{n\pi x}{L} \, dx \qquad\qquad n = 1, 2, 3, \ldots$$

Fourier Transform:

$$F(k) = \frac{1}{\sqrt{2\pi}} \int_{-\infty}^{+\infty} f(x')e^{-ikx'} dx' \qquad\qquad f(x) = \frac{1}{\sqrt{2\pi}} \int_{-\infty}^{+\infty} F(k')e^{ik'x} dk'$$

If $h(x) = af(x) + bg(x)$, $H(k) = aF(k) + bG(k)$ (linearity)	If $g(x) = f(x + \alpha)$, $G(k) = e^{+ik\alpha}F(k)$ (shifting)
If $g(x) = f(\alpha x)$, $G(k) = \dfrac{1}{\alpha} F\left(\dfrac{k}{\alpha}\right)$ (scaling)	If $g(x) = \dfrac{df}{dx}$, $G(k) = ikF(k)$ (differentiation)
If $h(x) = (f * g)(x)$, $H(k) = \sqrt{(2\pi)}F(k)G(k)$ (convolution in one domain is multiplication in the other)	

Special Waveforms, <u>all of period 2L</u>:

Square Wave, sine series

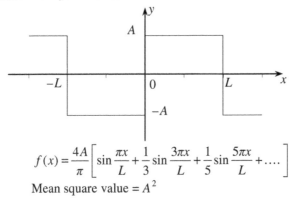

$$f(x) = \frac{4A}{\pi}\left[\sin\frac{\pi x}{L} + \frac{1}{3}\sin\frac{3\pi x}{L} + \frac{1}{5}\sin\frac{5\pi x}{L} + ...\right]$$

Mean square value $= A^2$

Square Wave, cosine series

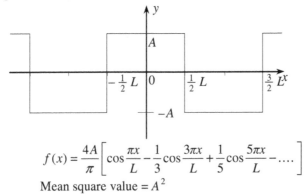

$$f(x) = \frac{4A}{\pi}\left[\cos\frac{\pi x}{L} - \frac{1}{3}\cos\frac{3\pi x}{L} + \frac{1}{5}\cos\frac{5\pi x}{L} - ....\right]$$

Mean square value $= A^2$

Triangular Wave

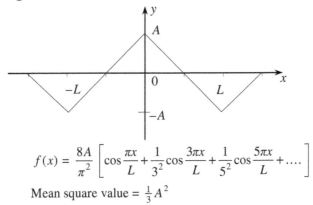

$$f(x) = \frac{8A}{\pi^2}\left[\cos\frac{\pi x}{L} + \frac{1}{3^2}\cos\frac{3\pi x}{L} + \frac{1}{5^2}\cos\frac{5\pi x}{L} + ...\right]$$

Mean square value $= \frac{1}{3}A^2$

3-6

Saw-Tooth Wave

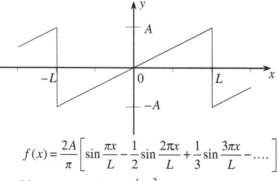

$$f(x) = \frac{2A}{\pi}\left[\sin\frac{\pi x}{L} - \frac{1}{2}\sin\frac{2\pi x}{L} + \frac{1}{3}\sin\frac{3\pi x}{L} - ....\right]$$

Mean square value $= \frac{1}{3}A^2$

Half-Wave Rectification

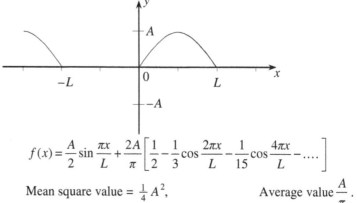

$$f(x) = \frac{A}{2}\sin\frac{\pi x}{L} + \frac{2A}{\pi}\left[\frac{1}{2} - \frac{1}{3}\cos\frac{2\pi x}{L} - \frac{1}{15}\cos\frac{4\pi x}{L} - ....\right]$$

Mean square value $= \frac{1}{4}A^2$,   Average value $\dfrac{A}{\pi}$.

Full-Wave Rectification

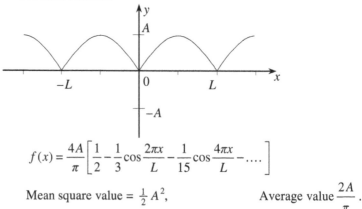

$$f(x) = \frac{4A}{\pi}\left[\frac{1}{2} - \frac{1}{3}\cos\frac{2\pi x}{L} - \frac{1}{15}\cos\frac{4\pi x}{L} - ....\right]$$

Mean square value $= \frac{1}{2}A^2$,   Average value $\dfrac{2A}{\pi}$.

3.3 *Trigonometric, Hyperbolic and Algebraic Relations*

Standard Trigonometrical Relationships:

$$
\begin{aligned}
\sin^2 A + \cos^2 A &= 1 \\
\sec^2 A &= \tan^2 A + 1 \\
2 \sin A \cos B &= \sin(A - B) + \sin(A + B) \\
2 \cos A \cos B &= \cos(A - B) + \cos(A + B) \\
2 \sin A \sin B &= \cos(A - B) - \cos(A + B) \\
\sin A + \sin B &= 2 \sin \tfrac{1}{2}(A + B) \cos \tfrac{1}{2}(A - B) \\
\sin A - \sin B &= 2 \cos \tfrac{1}{2}(A + B) \sin \tfrac{1}{2}(A - B) \\
\cos A + \cos B &= 2 \cos \tfrac{1}{2}(A + B) \cos \tfrac{1}{2}(A - B) \\
\cos A - \cos B &= - 2 \sin \tfrac{1}{2}(A + B) \sin \tfrac{1}{2}(A - B) \\
\sin(A \pm B) &= \sin A \cos B \pm \cos A \sin B \\
\cos(A \pm B) &= \cos A \cos B \mp \sin A \sin B \\
\tan(A \pm B) &= \frac{\tan A \pm \tan B}{1 \mp \tan A \tan B}
\end{aligned}
$$

$\tfrac{1}{2}\pi, 90°$

Positive Quadrants for Trigonometric Ratios:

$$
\begin{array}{c|c}
180° & \quad \sin \mid \text{all} \quad \quad 0 \\
\pi & \quad \tan \mid \cos \quad \quad 360°, 2\pi
\end{array}
$$

Relationships for Plane Triangle:

$$
\frac{a}{\sin A} = \frac{b}{\sin B} = \frac{c}{\sin C}
$$

$270°, \tfrac{3}{2}\pi$

$$
\sin A = \frac{2}{bc}\sqrt{s(s-a)(s-b)(s-c)}
$$

where $s = (a + b + c)$

$$
a^2 = b^2 + c^2 - 2bc \cos A
$$
$$
a^2 = b^2 + c^2 + 2bc \cos D
$$

Relationships for Spherical triangles

$$
\frac{\sin a}{\sin A} = \frac{\sin b}{\sin B} = \frac{\sin c}{\sin C}
$$

$$
\cos a = \cos b \cos c + \sin b \sin c \cos A
$$

$$
\cos A = -\cos B \cos C + \sin B \sin C \cos a
$$

$$
\sin \frac{A}{2} = \sqrt{\frac{\sin(s-b)\sin(s-c)}{\sin b \sin c}}
$$

where $s = \tfrac{1}{2}(a + b + c)$

$$
\sin \frac{a}{2} = \sqrt{-\frac{\cos S \cos(S - A)}{\sin B \sin C}}
$$

where $S = \tfrac{1}{2}(A + B + C)$

3-8

Napier's Rules for right spherical triangles:

Arrange the five parts about the right angle using the complement of the angle (co-x = $90° - x$) for the three parts not adjacent to the right angle. E.g. If the right angle is at A we have:

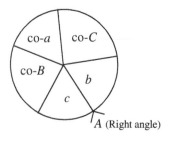

Then, the sine of the middle part is the product of the tangents of adjacent parts and is the product of the cosines of opposite parts.

N.B. A leg and its opposite angle are always in the same quadrant. If the hypotenuse is less than 90° the legs are in the same quadrant, otherwise they are in opposite quadrants.

Algebraic Relationships:

$$a^2 - b^2 = (a + b)(a - b) \qquad a^3 \pm b^3 = (a \pm b)(a^2 \mp ab + b^2)$$

$$a^x a^y = a^{x+y} \qquad (a^x)^y = a^{xy} \qquad a^0 = 1$$

Linear equation: $\quad y = mx + c \qquad y = 0$ at $x = -c/m$

Quadratic equation: $y = ax^2 + bx + c \qquad y = 0$ at $x = \dfrac{-b \pm \sqrt{b^2 - 4ac}}{2a}$

Logarithms:

$$x \equiv e^{\log_e x} \equiv e^{\ln x} \qquad\qquad e = 2.71828$$

$$x \equiv \log_{10}(10^x) \equiv \log_{10}(\text{antilog}_{10} x) \equiv 10^{\log_{10} x}$$

$$\ln x \equiv \log_e x = \frac{\log_{10} x}{\log_{10} e} = 2.30259 \log_{10} x$$

$$\log_{10} x = \log_{10} e \ln x = 0.4343 \ln x$$

Hyperbolics:

$$e^x = \cosh x + \sinh x \qquad\qquad \cosh^2 x - \sinh^2 x = 1$$

$$\sin x = \frac{e^{ix} - e^{-ix}}{2i} \qquad\qquad \cos x = \frac{e^{ix} + e^{-ix}}{2}$$

$$\sinh x = \frac{e^x - e^{-x}}{2} \qquad\qquad \cosh x = \frac{e^x + e^{-x}}{2}$$

$$\sin iz = i \sinh z \qquad\qquad \cos iz = \cosh z$$

$$\sinh iz = i \sin z \qquad\qquad \cosh iz = \cos z$$

3-9

Curves and their Equations:

Straight Line

$y = mx + c$

gradient m

Quadratic/Parabola

$y = ax^2 + bx + c$

Circle $x^2 + y^2 = a^2$

Ellipse $\dfrac{x^2}{a^2} + \dfrac{y^2}{b^2} = 1$

Hyperbola $\dfrac{x^2}{a^2} - \dfrac{y^2}{b^2} = 1$

Parabola $y^2 = ax$

Trigonometrical Functions

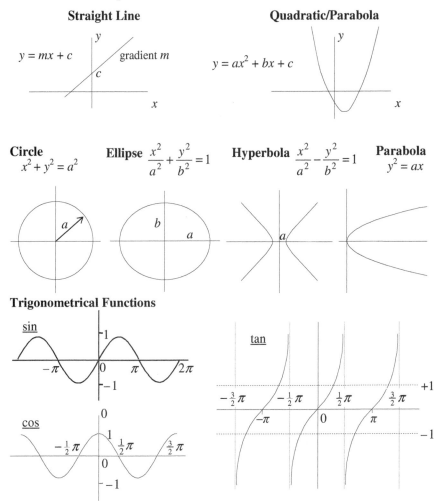

Exponential, Logarithmic and Hyperbolic Functions

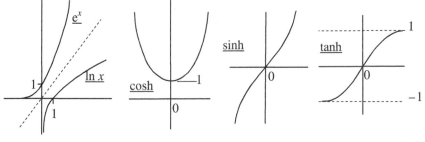

3.4 Complex Numbers

$$z = \quad x + i\,y \quad = \quad r\,(\cos\theta + i\sin\theta)$$
$$= \quad r\,e^{i(\theta + 2n\pi)} \qquad n = 0, \pm 1, \pm 2, \ldots$$
$$r = \sqrt{x^2 + y^2} \qquad \theta = \tan^{-1}\left(\frac{y}{x}\right)$$

N.B. θ must be chosen to lie in the appropriate quadrant – see diagram on page 3-8.

$$|z - z_0| \quad = \quad R \qquad \text{for circle}$$
$$z^c \quad = \quad e^{c\ln z}$$
$$e^{iz} \quad = \quad \cos z + i\sin z \qquad \text{(Euler's Formula)}$$

3.5 Partial Differentiation

If $\quad f = f(x,\,y)$, where $x = x(t)$ and $y = y(t)$, then

$$F = f(t) \text{ and } \quad \frac{df}{dt} = \frac{\partial f}{\partial x}\frac{dx}{dt} + \frac{\partial f}{\partial y}\frac{dy}{dt}$$

If $\quad f = f(x,\,y)$, where $y = y(x)$, then

$$F = f(x) \text{ and } \quad \frac{df}{dx} = \frac{\partial f}{\partial x} + \frac{\partial f}{\partial y}\frac{dy}{dx}$$

If $\quad f = f(x,\,y)$, $x = x(u,\,v)$ and $y = y(u,\,v)$, then

$$F = f(u,v) \text{ and } \quad \frac{df}{du} = \frac{\partial f}{\partial x}\frac{\partial x}{\partial u} + \frac{\partial f}{\partial y}\frac{\partial y}{\partial u}$$
$$\frac{df}{dv} = \frac{\partial f}{\partial x}\frac{\partial x}{\partial v} + \frac{\partial f}{\partial y}\frac{\partial y}{\partial v}$$

3.6 Differential Equations

First Order

Type	Characteristic	Solution Method
Separable	$y' = P(x)\,Q(y)$	$\int \frac{1}{Q}dy = \int P dx + c$
Homogeneous	$y' = f\left(\dfrac{y}{x}\right)$	substitution $y = ux$ makes equation separable
Exact	$M(x,\,y)dx + N(x,\,y)dy = 0$ where $\dfrac{\partial M}{\partial y} = \dfrac{\partial N}{\partial x}$	Solve for G where $\dfrac{\partial G}{\partial x} = M$, $\dfrac{\partial G}{\partial y} = N$
Linear	$y' + P(x)y = Q(x)$	multiply by $e^{\int P dx}$

Second Order, linear with constant coefficients

General Form:

$$\frac{d^2 y}{dx^2} + a\frac{dy}{dx} + by = f(x)$$

General Solution = Complementary Function + Particular Integral

or

Dynamic Response (GS) = Steady State (CF) + Transient Response (PI)

G.S. = C.F. + P.I.		
	Auxiliary equation: $m^2 + am + b = (m - m_1)(m - m_2) = 0$	
C.F. Transient	**Roots:** m_1 and m_2	**C.F.**
	m_1, m_2 real roots, $m_1 \neq m_2$	$A_1 e^{m_1 x} + A_2 e^{m_2 x}$
	$m_1 = m_2$ equal roots	$(A + Bx)e^{m_1 x}$
	m_1, m_2 complex conjugate roots $(\alpha \pm i\beta)$	$C e^{\alpha x} \cos(\beta x - \varepsilon)$
P.I. Steady State	**Forcing function:** $f(x)$	**P.I.**
	$f(x)$ is polynomial	Polynomial of same degree
	$f(x) = \lambda e^{\alpha x}$	
	$\quad \alpha \neq m_1$ or m_2	$Ae^{\alpha x}$
	$\quad \alpha = m_1 \neq m_2$	$Axe^{\alpha x}$
	$\quad \alpha = m_1 = m_2$	$Ax^2 e^{\alpha x}$
	$f(x) = F \cos \alpha x$	$A \cos(\alpha x + \phi)$
	$f(x) = F \sin \alpha x$	$A \sin(\alpha x + \phi)$

Transient behaviour: free vibration

$$m\ddot{x} + \alpha\dot{x} + kx = 0 \qquad \text{or} \qquad \ddot{x} + 2\xi\omega_0\dot{x} + \omega_0^2 x = 0$$

$$\text{where } \xi = \frac{\alpha}{2\sqrt{mk}} = \frac{\alpha}{2m\omega_0}, \quad \omega_0 = \sqrt{\frac{k}{m}}$$

$\xi < 1$ (underdamping)

$$x = a\,e^{-\xi\omega_0 t}\cos(\omega t - \theta)$$

where $\omega = \omega_0\sqrt{1-\xi^2}$

Ratio of successive peaks:

$$R = e^{\frac{2\pi\xi}{\sqrt{1-\xi^2}}}$$

Logarithmic decrement:

$$d = \ln R = \frac{2\pi\xi}{\sqrt{1-\zeta^2}} \approx 2\pi\xi \text{ for small } \xi.$$

$\xi > 1$ (overdamping)

$$x = A_1\,e^{-q_1 t} + A_2\,e^{-q_2 t}$$

where $q_1, q_2 = \omega_0(\xi \pm \sqrt{\xi^2 - 1})$

and $(q_1 > q_2)$

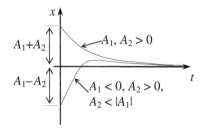

$\xi = 1$ (critical damping)

$$x = (A + Bt)e^{-\xi\omega_0 t}$$

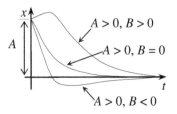

3-13

Constant amplitude forcing

$$m\ddot{x} + \alpha\dot{x} + kx = F\cos pt \quad \text{or} \quad \ddot{x} + 2\xi\omega_0\dot{x} + \omega_0^2 x = \frac{F}{m}\cos pt$$

$$\text{where } \xi = \frac{\alpha}{2\sqrt{mk}} = \frac{\alpha}{2m\omega_0}, \quad \omega_0 = \sqrt{\frac{k}{m}}$$

Particular integral is:

$$x = \frac{AF}{m\omega_0^2}\cos(pt - \phi) \quad \text{or} \quad x = Ax_1\cos(pt - \phi)$$

$$\text{where} \quad x_1 = \frac{F}{k}, \quad \tan\phi = \frac{2\xi\dfrac{p}{\omega_0}}{1 - \left(\dfrac{p}{\omega_0}\right)^2},$$

$$\text{and} \quad A = \left|\frac{x}{x_1}\right| = \frac{1}{\left[\left(1 - \left(\dfrac{p}{\omega_0}\right)^2\right)^2 + \left(2\xi\dfrac{p}{\omega_0}\right)^2\right]^{\frac{1}{2}}}$$

At resonance, $|x|$ is a maximum, and

$$\frac{x}{x_1} = \frac{1}{2\xi\sqrt{1-\xi^2}}, \quad p = \omega_0\sqrt{1-2\xi^2}, \quad \tan\phi = \frac{\sqrt{1-2\xi^2}}{\xi}$$

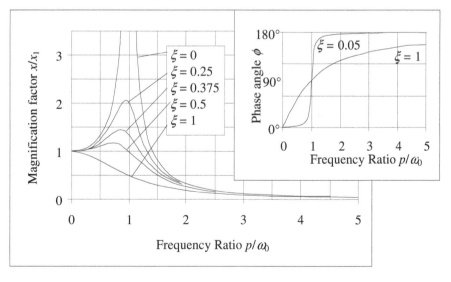

p^2 Forcing (out of balance mass etc.)

$$m\ddot{x} + \alpha\dot{x} + kx = mx_1 p^2 \cos pt \quad \text{or} \quad \ddot{x} + 2\xi\omega_0\dot{x} + \omega_0^2 x = x_1 p^2 \cos pt$$

$$\text{where} \quad \xi = \frac{\alpha}{2\sqrt{mk}} = \frac{\alpha}{2m\omega_0}, \quad \omega_0 = \sqrt{\frac{k}{m}}$$

Particular integral is:

$$x = Ax_1 \cos(pt - \phi)$$

$$\text{where} \quad \tan\phi = \frac{2\xi\dfrac{p}{\omega_0}}{1 - \left(\dfrac{p}{\omega_0}\right)^2}$$

$$\text{and} \quad A = \left|\frac{x}{x_1}\right| = \frac{\left(\dfrac{p}{\omega_0}\right)^2}{\left[\left(1 - \left(\dfrac{p}{\omega_0}\right)^2\right)^2 + \left(2\xi\dfrac{p}{\omega_0}\right)^2\right]^{\frac{1}{2}}}$$

At resonance, |x| is a maximum, and

$$\frac{x}{x_1} = \frac{1}{2\xi\sqrt{1-\xi^2}}, \quad p = \frac{\omega_0}{\sqrt{1-2\xi^2}}, \quad \tan\phi = \frac{-\sqrt{1-2\xi^2}}{\xi}$$

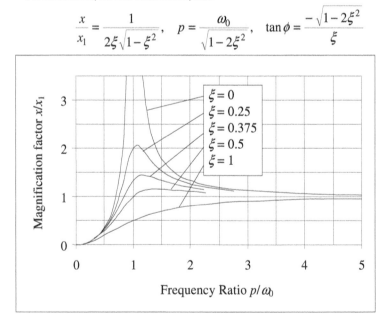

3.7 Rules for Differentiation and Integration

$\dfrac{dx}{dy} = \left(\dfrac{dy}{dx}\right)^{-1}$	$\dfrac{d}{dx}(f(x)+g(x)) = \dfrac{df(x)}{dx} + \dfrac{dg(x)}{dx}$
$\dfrac{d}{dx}(kf(x)) = k\dfrac{d}{dx}f(x)$	$\dfrac{d}{dx}(f(u)) = \dfrac{df}{du}\dfrac{du}{dx}$
$\dfrac{d}{dx}(uv) = u\dfrac{dv}{dx} + v\dfrac{du}{dx}$	$\dfrac{d}{dx}(uvw) = uv\dfrac{dw}{dx} + uw\dfrac{dv}{dx} + vw\dfrac{du}{dx}$
$\dfrac{d}{dx}\left(\dfrac{u}{v}\right) = \dfrac{1}{v^2}\left(v\dfrac{du}{dx} - u\dfrac{dv}{dx}\right)$	$\int uv\,dx = uw - \int \dfrac{du}{dx}w\,dx$, where $w = \int v\,dx$

3.8 Standard Differentials and Integrals

(Note: all the indefinite integrals below also require a constant of integration)

$$\frac{d}{dx}x^n = nx^{n-1}$$

$$\int x^n\,dx = \frac{x^{n+1}}{n+1}, \quad n \neq -1$$

$$\frac{d}{dx}\ln|x| = \frac{1}{x}$$

$$\int \frac{dx}{x} = \ln|x|$$

$$\frac{d}{dx}e^{ax} = ae^{ax}$$

$$\int e^{ax}\,dx = \frac{e^{ax}}{a}, \quad a \neq 0$$

$$\frac{d}{dx}a^x = a^x \ln a$$

$$\int a^x\,dx = \frac{a^x}{\ln a}, \quad a > 0, a \neq 1$$

$$\frac{d}{dx}x^x = x^x(1+\ln x)$$

$$\int \ln x\,dx = x(\ln x - 1)$$

$$\frac{d}{dx}\sin x = \cos x$$

$$\int \sin x\,dx = -\cos x$$

$$\frac{d}{dx}\cos x = -\sin x$$

$$\int \cos x\,dx = \sin x$$

$$\frac{d}{dx}\tan x = \sec^2 x$$

$$\int \tan x\,dx = -\ln|\cos x|$$

$$\frac{d}{dx}\operatorname{cosec} x = -\operatorname{cosec} x \cot x$$

$$\int \operatorname{cosec} x\,dx = \ln\left|\tan\frac{x}{2}\right|$$

$$\frac{d}{dx}\sec x = \sec x \tan x$$

$$\int \sec x\,dx = \ln\left|\tan\left(\frac{\pi}{4} + \frac{x}{2}\right)\right|$$

$$\frac{d}{dx}\cot x = -\operatorname{cosec}^2 x$$

$$\int \cot x\,dx = \ln|\sin x|$$

$$\frac{d}{dx}\sin^{-1} x = \frac{1}{\sqrt{1-x^2}}$$

$$\int \sec^2 x\,dx = \tan x$$

3-16

$$\frac{d}{dx}\cos^{-1}x = \frac{-1}{\sqrt{1-x^2}}$$

$$\int \text{cosec}^2 x\,dx = -\cot x$$

$$\frac{d}{dx}\tan^{-1}x = \frac{1}{1+x^2}$$

$$\int \frac{dx}{\sqrt{1-x^2}} = \sin^{-1}x,\quad |x|<1$$

$$\frac{d}{dx}\cot^{-1}x = -\frac{1}{1+x^2}$$

$$\int \frac{dx}{1+x^2} = \tan^{-1}x$$

$$\frac{d}{dx}\sinh x = \cosh x$$

$$\int \cosh x\,dx = \sinh x$$

$$\frac{d}{dx}\cosh x = \sinh x$$

$$\int \sinh x\,dx = \cosh x$$

$$\frac{d}{dx}\tanh x = \text{sech}^2 x$$

$$\int \text{sech}^2 x\,dx = \tanh x$$

$$\frac{d}{dx}\coth x = -\text{cosech}^2 x$$

$$\int \text{cosech}^2 x\,dx = -\coth x$$

$$\frac{d}{dx}\sinh^{-1}x = \frac{1}{\sqrt{x^2+1}}$$

$$\int \frac{dx}{\sqrt{x^2+a^2}} = \sinh^{-1}\frac{x}{a} = \ln\left|x+\sqrt{x^2+a^2}\right|$$

$$\frac{d}{dx}\cosh^{-1}x = \frac{1}{\sqrt{x^2-1}}$$

$$\int \frac{dx}{\sqrt{x^2-a^2}} = \cosh^{-1}\frac{x}{a} = \ln\left|x+\sqrt{x^2-a^2}\right|\Big|_{x\geq a}$$

$$\frac{d}{dx}\tanh^{-1}x = \frac{1}{1-x^2},\quad x^2<1$$

$$\int \frac{dx}{a^2-x^2} = \frac{1}{a}\tanh^{-1}\frac{x}{a} = \frac{1}{2a}\ln\left|\frac{a+x}{a-x}\right|\quad x^2<a^2$$

$$\frac{d}{dx}\coth^{-1}x = \frac{1}{1-x^2},\quad x^2>1$$

$$\int \frac{dx}{x^2-a^2} = -\frac{1}{a}\coth^{-1}\frac{x}{a} = \frac{1}{2a}\ln\left|\frac{x-a}{x+a}\right|\quad x^2>a^2$$

Some Definite Integrals (*m* and *n* are integers in the following)

$$\int_0^{\frac{\pi}{2}}\sin^n x\,dx = \int_0^{\frac{\pi}{2}}\cos^n x\,dx = \begin{cases} \dfrac{n-1}{n}\times\dfrac{n-3}{n-2}\cdots\dfrac{3}{4}\times\dfrac{1}{2}\times\dfrac{\pi}{2}, & n \text{ even} \\[2ex] \dfrac{n-1}{n}\times\dfrac{n-3}{n-2}\cdots\dfrac{4}{5}\times\dfrac{2}{3}\times 1, & n \text{ odd} \end{cases}$$

$$I_{m,n} = \int_0^{\frac{\pi}{2}}\sin^m x\cos^n x\,dx = \left(\frac{m-1}{m+n}\right)I_{m-2,n} = \left(\frac{n-1}{m+n}\right)I_{m,n-2},\quad m\neq -n$$

$$\int_0^{\pi}\sin mx\sin nx\,dx = \int_0^{\pi}\cos mx\cos nx\,dx = 0,\quad m\neq n$$

$$\int_0^\infty e^{-ax}\sin bx\,dx = \frac{b}{a^2+b^2} \quad (a>0) \qquad \int_0^\infty e^{-ax}\cos bx\,dx = \frac{a}{a^2+b^2} \quad (a>0)$$

$$\int_0^\pi \frac{\cos nx}{\cos x - \cos\phi}\,dx = \frac{\pi\sin n\phi}{\sin\phi} \qquad \int_0^\pi \sin nx\cos nx\,dx = 0$$

$$\int_0^\pi \frac{\sin x\cos x}{(1+a\cos x)^3}\,dx = \frac{-2a}{(1-a^2)^2} \qquad \int_0^{2\pi} \frac{dx}{(1+a\cos x)} = \frac{2\pi}{\sqrt{(1-a^2)}}$$

$$\int_0^\infty e^{-x^2}\,dx = \frac{\sqrt{\pi}}{2} \qquad \int_0^\pi \frac{\sin^2 x}{(1+a\cos x)^3}\,dx = \frac{\pi}{2(1-a^2)^{\frac{3}{2}}}$$

3.9 *Laplace Transforms*

Theorems: Definition: $F(s) = L[f(t)] = \displaystyle\int_0^\infty f(t)e^{-st}\,dt$

Linearity	$L[af(t)+bg(t)]$	$= aF(s)+bG(s)$
Final Value	$\lim_{t\to\infty} f(t)$	$= \lim_{s\to 0} sF(s)$
Initial Value	$\lim_{t\to 0} f(t)$	$= \lim_{s\to\infty} sF(s)$
Differentiation	$L\left[\dfrac{df(t)}{dt}\right]$	$= sF(s) - f(0)$
	$L\left[\dfrac{d^2 f(t)}{dt^2}\right]$	$= s^2 F(s) - sf(0) - \dfrac{df(0)}{dt}$
Integration	$L\left[\displaystyle\int f(t)dt\right]$	$= \dfrac{F(s)}{s} + \dfrac{f^{-1}(0)}{s}$
First Shifting	$L[e^{at}f(t)]$	$= F(s-a)$
Second Shifting	$L[f(t-a)]$	$= e^{-as}F(s) \qquad (t>a)$
Convolution: $L[f*g] \equiv L\left[\displaystyle\int_0^t f(u)g(t-u)du\right]$		$= F(s)G(s)$
Partial Differentiation	$L\left[\dfrac{\partial f(t,\alpha)}{\partial\alpha}\right]$	$= \dfrac{\partial}{\partial\alpha}F(s,\alpha)$
Time Multiplication	$L[tf(t)]$	$= -\dfrac{dF(s)}{ds}$

Transform Pairs

Function	Laplace Transform
1	$\dfrac{1}{s}$
step function: $\quad H(t-T) = 0 \qquad t < T$ $\qquad\qquad\qquad\quad = 1 \qquad t \geq T$	$\dfrac{1}{s}\,e^{-sT}$
t^n	$\dfrac{n!}{s^{n+1}}$
e^{-at}	$\dfrac{1}{s+a}$
$\sin \omega t$	$\dfrac{\omega}{s^2 + \omega^2}$
$\cos \omega t$	$\dfrac{s}{s^2 + \omega^2}$
$e^{-at} \sin \omega t$	$\dfrac{\omega}{(s+a)^2 + \omega^2}$
$e^{-at} \cos \omega t$	$\dfrac{s+a}{(s+a)^2 + \omega^2}$
$1 - e^{-t/T}$	$\dfrac{1}{s(1+Ts)}$
$\dfrac{\omega_n}{\sqrt{1-\xi^2}}\, e^{-\xi\omega_n t} \sin\!\left(\omega_n \sqrt{1-\xi^2}\, t\right)$	$\dfrac{1}{\left(1 + 2\dfrac{\xi s}{\omega_n} + \dfrac{s^2}{\omega_n^2}\right)}$
$1 - \dfrac{1}{\sqrt{1-\xi^2}}\, e^{-\xi\omega_n t} \sin\!\left(\omega_n \sqrt{1-\xi^2}\, t + \cos^{-1}\xi\right)$	$\dfrac{1}{s\left(1 + 2\dfrac{\xi s}{\omega_n} + \dfrac{s^2}{\omega_n^2}\right)}$

See also z-Transforms, section 10.6.

3.10 *The Error Function – Selected Values:*

Definition: $\quad \text{erf } z = \dfrac{2}{\sqrt{\pi}} \displaystyle\int_0^z e^{-u^2}\, du$

z	erf z	z	erf z	z	erf z	z	erf z
0.00	0.0000	0.62	0.6194	1.24	0.9205	1.86	0.9915
0.02	0.0226	0.64	0.6346	1.26	0.9252	1.88	0.9922
0.04	0.0451	0.66	0.6494	1.28	0.9297	1.90	0.9928
0.06	0.0676	0.68	0.6638	1.30	0.9340	1.92	0.9934
0.08	0.0901	0.70	0.6778	1.32	0.9381	1.94	0.9939
0.10	0.1125	0.72	0.6914	1.34	0.9419	1.96	0.9944
0.12	0.1348	0.74	0.7047	1.36	0.9456	1.98	0.9949
0.14	0.1569	0.76	0.7175	1.38	0.9490	2.00	0.9953
0.16	0.1790	0.78	0.7300	1.40	0.9523	2.05	0.9963
0.18	0.2009	0.80	0.7421	1.42	0.9554	2.10	0.9970
0.20	0.2227	0.82	0.7538	1.44	0.9583	2.15	0.9976
0.22	0.2443	0.84	0.7651	1.46	0.9611	2.20	0.9981
0.24	0.2657	0.86	0.7761	1.48	0.9637	2.25	0.9985
0.26	0.2869	0.88	0.7867	1.50	0.9661	2.30	0.9989
0.28	0.3079	0.90	0.7969	1.52	0.9684	2.35	0.9991
0.30	0.3286	0.92	0.8068	1.54	0.9706	2.40	0.9993
0.32	0.3491	0.94	0.8163	1.56	0.9726	2.45	0.9995
0.34	0.3694	0.96	0.8254	1.58	0.9745	2.50	0.9996
0.36	0.3893	0.98	0.8342	1.60	0.9763	2.55	0.9997
0.38	0.4090	1.00	0.8427	1.62	0.9780	2.60	0.9998
0.40	0.4284	1.02	0.8508	1.64	0.9796	2.65	0.9998
0.42	0.4475	1.04	0.8586	1.66	0.9811	2.70	0.9999
0.44	0.4662	1.06	0.8661	1.68	0.9825	2.75	0.9999
0.46	0.4847	1.08	0.8733	1.70	0.9838	2.80	0.9999
0.48	0.5027	1.10	0.8802	1.72	0.9850	2.85	0.9999
0.50	0.5205	1.12	0.8868	1.74	0.9861	2.90	1.0000
0.52	0.5379	1.14	0.8931	1.76	0.9872	2.95	1.0000
0.54	0.5549	1.16	0.8991	1.78	0.9882	3.00	1.0000
0.56	0.5716	1.18	0.9048	1.80	0.9891	4.00	1.0000
0.58	0.5879	1.20	0.9103	1.82	0.9899		
0.60	0.6039	1.22	0.9155	1.84	0.9907		

3.11 *Numerical Analysis*

Approximate solution of an Algebraic Equation $f(x) = 0$

Newton's Method

$$x_1 = x_0 - \frac{f(x_0)}{f'(x_0)}$$

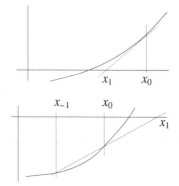

Secant Method

$$x_1 = \frac{-x_0 f(x_{-1}) + x_{-1} f(x_0)}{f(x_0) - f(x_{-1})}$$

Least Squares curve fitting of a straight line

If y_i $(i = 1, 2, \ldots n)$ are experimental values of y at chosen exact values of x_i, then the line of best fit passes through the centroid of the points:

$$\bar{x} = \frac{1}{n}\sum_{i=1}^{n} x_i \qquad \bar{y} = \frac{1}{n}\sum_{i=1}^{n} y_i$$

and is given by $y = mx + c$, where

$$m = \frac{\sum (x_i - \bar{x})(y_i - \bar{y})}{\sum (x_i - \bar{x})^2} = \frac{\sum x_i y_i - n\bar{x}\bar{y}}{\sum x_i^2 - n\bar{x}^2}$$

$$c = \bar{y} - m\bar{x}$$

Finite Difference Formulae

$$\Delta f(x) = f(x + h) - f(x)$$

$$f'(x) = \frac{f(x+h) - f(x-h)}{2h} + O(h^2)$$

$$f''(x) = \frac{f(x+h) - 2f(x) + f(x-h)}{h^2} + O(h^2)$$

$$f'''(x) = \frac{f(x+2h) - 2f(x+h) + 2f(x-h) - f(x-2h)}{2h^3}$$

Lagrange's Interpolation Formula for unequal intervals

The polynomial $P(x)$ of degree 2 passing through three points (x_i, y_i), $i = 1,2,3$, is

$$P(x) = \frac{(x - x_2)(x - x_3)}{(x_1 - x_2)(x_1 - x_3)} y_1 + \frac{(x - x_1)(x - x_3)}{(x_2 - x_1)(x_2 - x_3)} y_2 + \frac{(x - x_1)(x - x_2)}{(x_3 - x_1)(x_3 - x_2)} y_3$$

3-21

Formulae for Numerical Integration

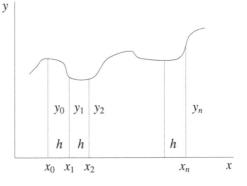

Divide the range into
equal intervals of width
h such that

$x_n = x_0 + nh$

$y_n = y_n(x_n)$

Trapezoidal Rule:

1 strip:
$$\int_{x_0}^{x_1} y(x)dx = \frac{h}{2}[y_0 + y_1] + \varepsilon, \quad \text{where } \varepsilon \approx \frac{-h^3}{12} y_0''$$

n strips:
$$\int_{x_0}^{x_n} y(x)dx \approx \frac{h}{2}[y_0 + 2y_1 + 2y_2 + ... + 2y_{n-1} + y_n]$$

Simpson's Rule:

2 strips:
$$\int_{x_0}^{x_2} y(x)dx = \frac{h}{3}[y_0 + 4y_1 + y_2] + \varepsilon, \quad \text{where } \varepsilon \approx \frac{-h^5}{90} y_1^{(4)}$$

n strips (n even):
$$\int_{x_0}^{x_n} y(x)dx \approx \frac{h}{3}[y_0 + y_n + 4(y_1 + y_3 + ... + y_{n-1}) + 2(y_2 + y_4 + ... + y_{n-2})]$$

$$= (h/3) \times (\text{"first + last + four} \times \text{odds + two} \times \text{evens"})$$

Runge-Kutta Method:

Second Order:
$$y_{n+1} = y_n + \frac{h}{2}\{f(x_n, y_n) + f(x_n + h, y_n + k_1)\}$$

Fourth Order:
$$y_{n+1} = y_n + \frac{1}{6}(k_1 + 2k_2 + 2k_3 + k_4)$$

$$k_1 = hf(x_n, y_n)$$

$$k_2 = hf\left(x_n + \frac{h}{2}, y_n + \frac{k_1}{2}\right)$$

$$k_3 = hf\left(x_n + \frac{h}{2}, y_n + \frac{k_2}{2}\right)$$

$$k_4 = hf(x_n + h, y_n + k_3)$$

4. Analysis of Experimental Data

4.1 *Probability Distributions for Discrete Random Variables*

Notation:

$P(r) = f(r) \Rightarrow$ the probability distribution of random variable r is $f(r)$

μ = mean value of r = $\displaystyle\sum_{i=1}^{n} r_i f(r_i)$

σ^2 = variance of r = $\displaystyle\sum_{i=1}^{n} r_i^2 f(r_i) - \mu^2$

σ = population standard deviation of r

$\dbinom{n}{r}$ = binomial coefficient = $\dfrac{n!}{(n-r)!\,r!} = \dbinom{n}{n-r}$

Evaluate using Pascal's Triangle:

	r = 0	1	2	3	4	5	6	7	8	9	10
n = 0	1										
1	1	1									
2	1	2	1								
3	1	3	3	1							
4	1	4	6	4	1						
5	1	5	10	10	5	1					
6	1	6	15	20	15	6	1				
7	1	7	21	35	35	21	7	1			
8	1	8	28	56	70	56	28	8	1		
9	1	9	36	84	126	126	84	36	9	1	
10	1	10	45	120	210	252	210	120	45	10	1

Binomial Distribution

n = number of trials with constant probability p of success in each
r = number of successes

$$P(r) = \binom{n}{r} p^r (1-p)^{n-r} \qquad r = 0, 1, 2, \ldots n$$

$$\mu = np \qquad\qquad \sigma^2 = np(1-p)$$

Poisson Distribution

μ = mean rate of occurrence of an event
r = number of events actually occurring in unit time

$$P(r) = e^{-\mu}\frac{\mu^r}{r!} \qquad r = 0, 1, \ldots$$

$$\sigma^2 = \mu$$

4.2 *Probability Distributions for Continuous Random Variables*

Exponential Distribution

probability density function $\qquad f(x) = \lambda e^{-\lambda x} \qquad x \geq 0,\ \lambda > 0$

$$\mu = 1/\lambda \qquad\qquad \sigma^2 = 1/\lambda^2$$

Normal Distribution

The *standardised* normal distribution $N(0, 1)$ has $\mu = 0$ and $\sigma = 1$. Its probability density function is:

$$\phi(z) = \frac{1}{\sqrt{2\pi}} e^{-\frac{1}{2}z^2}$$

Its cumulative distribution function $\Phi(z) = \displaystyle\int_{-\infty}^{z} \frac{1}{\sqrt{2\pi}} e^{-\frac{1}{2}t^2}\, dt$

is the probability that the random variable is observed to have a value $\leq z$ (the shaded area shown in the graph of $N(0,1)$ on page 4-3).

Percentage Points of the Normal Distribution $N(0, 1)$			
$\Phi(z)$	% (1 tail)	% (2 tails)	z
0.9500	5.0	10	1.6449
0.9750	2.5	5	1.9600
0.9900	1.0	2	2.3263
0.9950	0.5	1	2.5758

Cumulative Distribution Function for $N(0, 1)$					
z	$\Phi(z)$	z	$\Phi(z)$	z	$\Phi(z)$
0.0	0.5000	1.0	0.8413	2.0	0.9772
0.1	0.5398	1.1	0.8643	2.1	0.9821
0.2	0.5793	1.2	0.8849	2.2	0.9861
0.3	0.6179	1.3	0.9032	2.3	0.9893
0.4	0.6554	1.4	0.9192	2.4	0.9918
0.5	0.6915	1.5	0.9332	2.5	0.9938
0.6	0.7257	1.6	0.9452	2.6	0.9953
0.7	0.7580	1.7	0.9554	2.7	0.9965
0.8	0.7881	1.8	0.9641	2.8	0.9974
0.9	0.8159	1.9	0.9713	2.9	0.9981
For negative values of z, use $\Phi(-z) = 1 - \Phi(z)$				3.0	0.9987
				4.0	>0.9999

The *general* normal distribution $N(\mu, \sigma^2)$ has probability density function

$$f(x) = \frac{1}{\sigma\sqrt{2\pi}} e^{-\frac{(x-\mu)^2}{2\sigma^2}} = \phi\left(\frac{x-\mu}{\sigma}\right), \text{ where } \int_{-\infty}^{\infty} f(x)dx = 1$$

Its cumulative distribution $F(x) = \displaystyle\int_{-\infty}^{(x-\mu)/\sigma} \frac{1}{\sqrt{2\pi}} e^{-\frac{1}{2}t^2} \, dt = \Phi\left(\frac{x-\mu}{\sigma}\right).$

Tables of $\phi(z)$ and $\Phi(z)$ may be used for $f(x)$ and $F(x)$ by taking

$z = \dfrac{x-\mu}{\sigma}.$

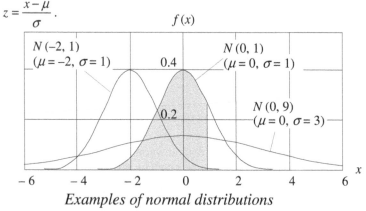

Examples of normal distributions

4.3 *Experimental Samples*

$x_1, x_2, \ldots x_n$ denote a set of n observations of a random variable having a normal distribution whose mean μ is unknown.

Range:	$x_{max} - x_{min}$
Sample mean:	$m = \dfrac{1}{n}\sum x_i$
Average deviation:	$\dfrac{1}{n}\sum |x_i - m|$
Sample variance:	$s^2 = \dfrac{1}{n-1}\sum (x_i - m)^2$
Sample Standard Deviation:	s
Distribution of x:	$N(\mu, \sigma^2)$
Distribution of m:	$N(\mu, \sigma^2/n)$
Distribution of $\dfrac{m-\mu}{(\sigma/\sqrt{n})}$	$N(0, 1)$
Standard error of sample means:	$S = \dfrac{\sigma}{\sqrt{n}}$

If the population variance σ^2 is known,

95% confidence interval for μ is $\quad m \pm 1.96 \dfrac{\sigma}{\sqrt{n}}$

$$99\% \text{ confidence interval for } \mu \text{ is } \quad m \pm 2.58 \frac{\sigma}{\sqrt{n}}$$

If the population variance σ^2 is unknown, $\dfrac{m - \mu}{s/\sqrt{n}}$ has the t distribution
with $n - 1$ degrees of freedom $(t_n - 1)$, and the 95% confidence interval
for μ is obtained from $m \pm t_c \dfrac{s}{\sqrt{n}}$ and the table.

95% points of the t-distribution					
$n - 1$	t_c	$n - 1$	t_c	$n - 1$	t_c
1	12.7	6	2.45	12	2.18
2	4.30	7	2.36	15	2.13
3	3.18	8	2.31	20	2.09
4	2.78	9	2.26	30	2.04
5	2.57	10	2.23	60	2.00
				∞	1.96

Thus, for $n > 20$, $m \pm 1.96\, s/\sqrt{n}$ is a good approximation to the
population mean with a 95% confidence.

4.4 *Combination of Errors*

If results are normally distributed, the Most Probable Error S_z in the
calculated result $z = f(x, y, \text{etc})$ due to independent standard errors S_x, S_y,
etc in x, y, etc is given by:

$$(S_z)^2 = \left(\frac{\partial z}{\partial x} S_x \right)^2 + \left(\frac{\partial z}{\partial y} S_y \right)^2 + \dots \text{etc}$$

If the function $f(\ )$ consists of multiplied and divided terms *only* (i.e. no
addition or subtraction):

$$\left(\frac{S_z}{z} \right)^2 = \left(n \frac{S_x}{x} \right)^2 + \left(m \frac{S_y}{y} \right)^2 + \dots \text{etc}$$

where n, m etc are the powers of x, y etc in $f(\ )$.
The standard error S_d of the difference between two mean values having
standard errors S_1 and S_2 is $S_d = \sqrt{S_1^2 + S_2^2}$

Notes:

(a)   The maximum possible error $\left(S_z = \dfrac{\partial z}{\partial x} S_x + \dfrac{\partial z}{\partial y} S_y + \dots \text{etc} \right)$ is
       rarely of interest in engineering.

(b)   Instrument 'rounding off' error $\pm \delta x$ may be treated as a
       normally distributed error by the equivalence $S_x \approx \frac{2}{3} \delta x$.

5. Mechanics

5.1 *Second Moment of Area*

Definition:

$$I_{XX} = \int\int (y - y_1)^2 \, dx \, dy$$

The double integral is taken over the whole area of the shape. Similar expressions apply to axes in other directions.

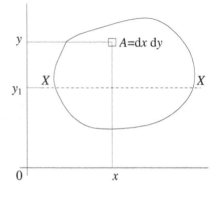

Radius of gyration k is defined by $I = Ak^2$ for a specified axis.

Parallel axis theorem:

$$I_{XX} = I_{GG} + A \, y_0^2$$

where A is the area of the shape and y_0 is the perpendicular distance between an axis through the centroid, GG, and another parallel axis XX. I_{GG} is the minimum second moment of area for all axes parallel to GG.

Radius of Gyration of selected shapes

Shape	k_{XX}^2	k_{YY}^2	A
Rectangle	$\dfrac{1}{12}d^2$	$\dfrac{1}{12}b^2$	db
Circle	$\dfrac{1}{4}r^2$	$\dfrac{1}{4}r^2$	πr^2

Shape	k_{XX}^2	k_{YY}^2	A
Ring 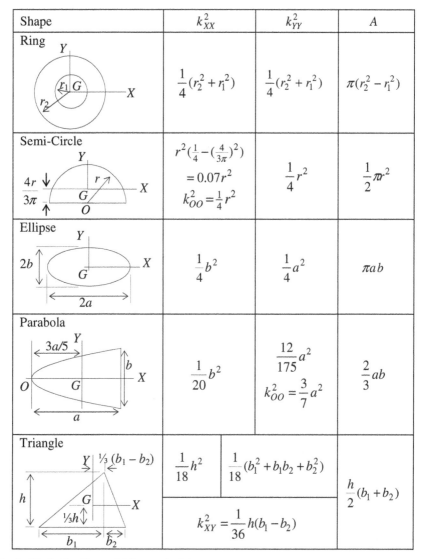	$\frac{1}{4}(r_2^2 + r_1^2)$	$\frac{1}{4}(r_2^2 + r_1^2)$	$\pi(r_2^2 - r_1^2)$
Semi-Circle	$r^2(\frac{1}{4} - (\frac{4}{3\pi})^2)$ $= 0.07r^2$ $k_{OO}^2 = \frac{1}{4}r^2$	$\frac{1}{4}r^2$	$\frac{1}{2}\pi r^2$
Ellipse	$\frac{1}{4}b^2$	$\frac{1}{4}a^2$	$\pi a b$
Parabola	$\frac{1}{20}b^2$	$\frac{12}{175}a^2$ $k_{OO}^2 = \frac{3}{7}a^2$	$\frac{2}{3}ab$
Triangle	$\frac{1}{18}h^2$	$\frac{1}{18}(b_1^2 + b_1 b_2 + b_2^2)$	$\frac{h}{2}(b_1 + b_2)$
	$k_{XY}^2 = \frac{1}{36}h(b_1 - b_2)$		

Mohr's Circle for Second Moment of Area

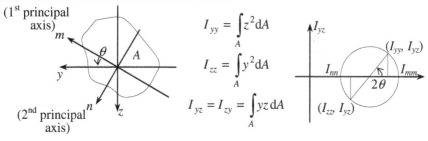

$$I_{yy} = \int_A z^2 \, \mathrm{d}A$$

$$I_{zz} = \int_A y^2 \, \mathrm{d}A$$

$$I_{yz} = I_{zy} = \int_A yz \, \mathrm{d}A$$

5.2 *Kinematics and Dynamics*

Constant Acceleration Equations

$$v = u + at \qquad\qquad v^2 = u^2 + 2ax$$

$$x = ut + \tfrac{1}{2}at^2 \qquad\qquad x = \tfrac{1}{2}(u + v)t$$

Accelerations due to Rotation

Central (centripetal) $\qquad = \qquad \underline{\omega}\times(\underline{\omega}\times\underline{r}) = r\omega^2 = \dfrac{v^2}{r}$

Coriolis $\qquad\qquad\qquad = \qquad 2\underline{\omega}\times\dfrac{d\underline{r}}{dt}$

Force, Work and Energy

	Linear	*Rotational*
Newton's 2nd Law	$F = ma$	$T = I\ddot{\theta} = I\dot{\omega}$
Momentum, Impulse	$\int F dt = mv$	
Angular Momentum		$\int T dt = I\dot{\theta} = I\omega$
Work, energy	$\int F dx$	$\int T d\theta$
Kinetic Energy	$\tfrac{1}{2}mv^2$	$\tfrac{1}{2}I\dot{\theta}^2 = \tfrac{1}{2}I\omega^2$
Power	Fv	$T\dot{\theta} = T\omega$

Lagrange's Equation

$$\frac{d}{dt}\left\{\frac{\partial\,\text{K.E.}}{\partial\dot{q}_i}\right\} - \frac{\partial\,\text{K.E.}}{\partial q_i} + \frac{\partial\,\text{P.E.}}{\partial q_i} + \frac{\partial\,\text{D.E.}}{\partial q_i} = Q_i$$

$i = 1,2,3$ for a 3 degree of freedom system.
q_i is a generalised co-ordinate
Q_i is a generalised force
K.E., P.E., D.E. are kinetic, potential and dissipative energy.

Friction:

Coefficient of static friction:

For no slipping: $\quad \mu = \tan\phi \geq \dfrac{F}{N}$

(ϕ = angle of friction)

Around drum or pulley: $\quad \dfrac{F_1}{F_2} = e^{\mu\theta}$

Approximate Values of Friction Coefficients:

Material Combination	μ Static		μ Dynamic	
	Dry	Lubricated	Dry	Lubricated
Steel/cast iron	0.2	0.1	0.18	0.06
Leather/metal	0.6	-	0.48	0.15
Brake or clutch lining/steel	-	-	0.5	-
Rubber/ashphalt	0.8	-	0.5	-
Filled PTFE/steel	-	-	0.05-0.3	-

5.3 *Moments of Inertia*

Definitions:

$$\text{Moment of Inertia: } I_{XX} = \int_M r^2 dm \text{ ; Radius of gyration: } k_{XX} = \sqrt{\frac{I_{XX}}{M}}$$

r is the perpendicular distance of the element of mass dm from the axis XX; the integral is taken over the whole mass M of the body.

Parallel axis theorem:

$$I_{XX} = I_{GG} + M\, y_0^2$$

M is the mass of the body, y_0 is the perpendicular distance between an axis GG through the centre of mass and another axis XX parallel to GG. I_{GG} is the minimum moment of inertia for all axes in the given direction.

Moment of inertia of a lamina

For a lamina with uniform mass distribution over its area, the moment of inertia about any axis XX in its plane is:

$$I_{XX} = (\text{mass per unit area}) \times (\text{second moment of area about } XX).$$

Perpendicular Axes Theorem

$$J = I_{ZZ} = I_{XX} + I_{YY}$$

I_{XX} and I_{YY} are the moments of inertia of a lamina about any two perpendicular axes in its plane.
J is the Polar Moment of Inertia. $J = I_{ZZ}$, the moment of inertia about a third axis ZZ, perpendicular to and intersecting both XX and YY.

Moment of Inertia of a Uniform Density Body

Axially-symmetric bodies may be treated as a succession of laminae.

Shape	k_{XX}^2	k_{YY}^2	Volume
Rod	-	$\dfrac{1}{12}l^2$	-
Solid Cylinder	$\dfrac{1}{2}r^2$	$\dfrac{1}{4}r^2 + \dfrac{1}{12}l^2$	$\pi r^2 l$
Thin-walled Cylinder	$r^2 + \dfrac{1}{4}t^2$	$\dfrac{1}{2}r^2 + \dfrac{1}{8}t^2 + \dfrac{1}{12}l^2$	$2\pi r t l$
Thick-walled Cylinder	$\dfrac{1}{2}(R^2 + r^2)$	$\dfrac{1}{12}l^2 + \dfrac{1}{4}(R^2 + r^2)$	$\pi(R^2 - r^2)l$
Sphere	$\dfrac{2}{5}r^2$	$\dfrac{2}{5}r^2$	$\dfrac{4}{3}\pi r^3$
Cone	$\dfrac{3}{80}(4r^2 + h^2)$	$\dfrac{3}{10}r^2$	$\dfrac{1}{3}\pi r^2 h$
Rectangular Block	$\dfrac{1}{12}(d^2 + t^2)$	$\dfrac{1}{12}(b^2 + t^2)$	bdt

6. Properties and Mechanics of Solids

6.1 *Bonding*

Condon-Morse Equation $\quad V_{total} = -\dfrac{Ae^2}{r^n} + \dfrac{B}{r^m} + C$

Ionic Bond Equation $\quad V_0 = -\dfrac{Z_1^* Z_2^* e^2}{4\pi d \varepsilon_0}(n - \dfrac{1}{n}) + \Delta E$

Theoretical Density $\quad \rho = \dfrac{nA}{VN}$

6.2 *Atomic sizes in substitutional alloys*

Element	Seitz radius r_0 (nm) at 20°C	Effective Valency in solution	Element	Seitz radius r_0 (nm) at 20°C	Effective Valency in solution
Al	0.158	3	P	0.158	3
Au	0.159	1	Pb	0.195	4
Cu	0.141	1	Si	0.167	4
Fe(α)	0.141	?	Sn	0.186	4
Mg	0.185	2	Zn	0.154	2
Ni	0.138	1			

6.3 *Phase Transformations*

Length and volume changes may be related by:

$$(1 + \Delta V/V) = (1 + \Delta L/L)^3$$

6.4 *Crystallography*

In the Miller index system:

Specific plane	(h.k.l)
Family of planes	{h.k.l}
Specific direction	[h.k.l]
Family of directions	<h.k.l>

Inter-planar spacing for cubics: $\quad d_{(h.k.l)} = \dfrac{a}{\sqrt{h^2 + k^2 + l^2}} = \dfrac{a}{\sqrt{N}}$

Quadratic form of Miller Indices for Cubic Structures

	N values
Simple	1,2,3,4,5,6,7,8,9,10,11,12,13,14,15,16,17,18,19,20,...
Face centred	3,4,8,11,12,16,19,20,24,27,32,...
Body centred	2,4,6,8,10,12,14,16,18,20,22,24,26,30,...
Diamond	3,8,11,16,19,24,27,32,...

6.5 Defects and Diffusion Data

Number of defects

$$n = \alpha N' e^{-\frac{Q}{kT}}$$

Diffusivity

$$D = D_0 e^{-\frac{Q}{kT}}$$

(Note: these equations may be expressed in terms of R_0 rather than k, but the value of Q must be in the appropriate units.)

Macroscopic Diffusion:

(i) D constant with composition, $\dfrac{dC}{dx}$ constant with time

(special case of (ii) below):

$$J = -D\frac{dC}{dx}$$

(ii) D constant with composition, $\dfrac{dC}{dx}$ varies with time:

$$\frac{dC}{dt} = D\frac{d^2C}{dx^2}$$

Solution for case (ii) with constant surface concentration and impermeable sides:

$$\frac{C_x - C_0}{C_s - C_0} = 1 - \mathrm{erf}\left(\frac{x}{2\sqrt{Dt}}\right)$$

(Values of the error function erf( ) are tabulated on page 3-20.)

6.6 *Fracture*

Fracture Toughness:

General Stress Intensity $\quad K_1 = \sigma\sqrt{a}\ \gamma(\dfrac{a}{W})$

Crack configurations with corresponding values of $\gamma(\dfrac{a}{W})$

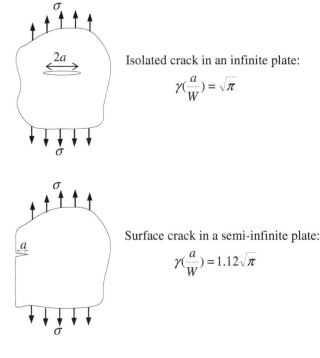

Isolated crack in an infinite plate:

$$\gamma(\frac{a}{W}) = \sqrt{\pi}$$

Surface crack in a semi-infinite plate:

$$\gamma(\frac{a}{W}) = 1.12\sqrt{\pi}$$

Surface crack in a three-point bend specimen:

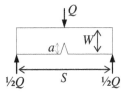

$$\sigma = \frac{6M}{tW^2}, \quad M = \frac{QS}{4}$$

where t is the plate thickness

$$\gamma(\frac{a}{W}) = 1.93 - 3.07\left(\frac{a}{W}\right) + 14.53\left(\frac{a}{W}\right)^2 - 25.11\left(\frac{a}{W}\right)^3 + 25.8\left(\frac{a}{W}\right)^4$$

Fatigue:

Manson-Coffin Law $\qquad \sqrt{N_p}\,\varepsilon_p = c = \text{constant}$

Miner's Rule $\qquad \sum \left(\dfrac{n_i}{N_i} \right) = 1$

Rayleigh Distribution $\qquad P(\sigma) = \sigma\, r^{-2} \exp\left\{ -\dfrac{1}{2}(\dfrac{\sigma}{r})^2 \right\}$

Fraction of peaks
exceeding stress (σ) $\qquad E(\sigma) = \exp\left\{ -\dfrac{1}{2}(\dfrac{\sigma}{r})^2 \right\}$
expressed in terms of E

Paris Equation $\qquad \dfrac{da}{dN} = A(\Delta K)^m = A_1 a^{m/2}$ (A, A_1 and m are constants)

Fatigue crack growth data for various materials

Material	R $\sigma_{min}/\sigma_{max}$	m	ΔK for $\dfrac{da}{dN} = 10^{-6}\,\text{mm/c}$ $(\text{MN/m}^{3/2})$
Mild steel	0.06 - 0.74	3.3	6.2
in brine	0.64	3.3	6.2
cold rolled	0.07 - 0.43	4.2	7.2
Low alloy steel	0 - 0.75	3.3	5.1
Maraging steel	0.67	3.0	3.5
Austenitic steel 18/8	0.33 - 0.43	3.1	6.3
Aluminium	0.14 - 0.87	2.9	2.9
Aluminium alloy 5083	0.20 - 0.69	2.7	1.6
Aluminium alloy 6082 1% Mg, 1% Si, 0.7% Mn	0.20 - 0.71	2.6	1.9
Aluminium alloy L73 4.5% Cu	0.50 - 0.88	4.4	2.1
Aluminium alloy DTD 687A 5.5% Zn	0.20 - 0.45	3.7	1.75
Magnesium alloy ZW1 0.5% Zr	0	3.35	0.94
Magnesium alloy AM503 1.5% Mn	0.5	3.35	0.69
Copper	0.07 - 0.82	3.9	4.3
Phosphor bronze	0.33 - 0.74	3.9	4.3
Brass 60/40	0 - 0.33	4.0	6.3
Titanium	0.08 - 0.94	4.4	3.1
Titanium alloy 5%Al	0.17 - 0.86	3.8	3.4
Titanium alloy 15% Mo	0.28 - 0.71	3.5	3.0
Nickel	0 - 0.71	4.0	8.8
Monel	0 - 0.67	4.0	6.2
Inconel	0 - 0.71	4.0	8.2

6.7 *Corrosion Potentials*

Standard oxidation/reduction (redox) potentials

at 25°C, against normal hydrogen electrode

	volts
$Au = Au^{3+} + 3e$	+1.498
$O_2 + 4H^+ + 4e = 2H_2O$	+1.229
$Pt = Pt^{2+} + 2e$	+1.2
$Pd = Pd^{2+} + 2e$	+0.987
$Ag = Ag^+ + e$	+0.799
$2Hg = Hg_2^{2+} + 2e$	+0.788
$Fe^{3+} + e = Fe^{2+}$	+0.771
$O_2 + 2H_2O + 4e = 4\,OH^-$	+0.401
$Cu = Cu^{2+} + 2e$	+0.337
$Sn^{4+} + 2e = Sn^{2+}$	+0.15
$2H^+ + 2e = H_2$	0.000
$Pb = Pb^{2+} + 2e$	−0.126
$Sn = Sn^{2+} + 2e$	−0.136
$Ni = Ni^{2+} + 2e$	−0.250
$Co = Co^{2+} + 2e$	−0.277
$Cd = Cd^{2+} + 2e$	−0.403
$Fe = Fe^{2+} + 2e$	−0.440
$Cr = Cr^{3+} + 3e$	−0.744
$Zn = Zn^{2+} + 2e$	−0.763
$Al = Al^{3+} + 3e$	−1.662
$Mg = Mg^{2+} + 2e$	−2.363
$Na = Na^+ + e$	−2.714
$K = K^+ + e$	−2.925

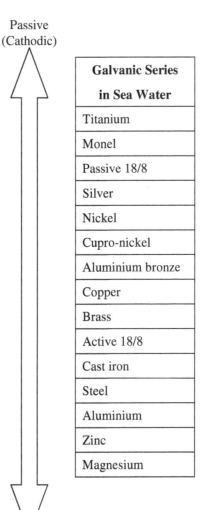

Passive
(Cathodic)

Galvanic Series

in Sea Water

Titanium

Monel

Passive 18/8

Silver

Nickel

Cupro-nickel

Aluminium bronze

Copper

Brass

Active 18/8

Cast iron

Steel

Aluminium

Zinc

Magnesium

Active
(Anodic)

6.8 *Structure and Properties of some Elements*

	Metals			Semiconductors	
	Copper Cu	Iron Fe	Aluminium Al	Germanium Ge	Silicon Si
Crystal Structure	f.c.c.	b.c.c.	f.c.c.	diamond	diamond
Bonding	metallic	metallic	metallic	covalent	covalent
Lattice constant (nm)	0.3615	0.2866	0.4049	0.5658	0.5431
Atomic volume (nm^3)	11.81×10^{-3}	11.77×10^{-3}	16.6×10^{-3}	22.64×10^{-3}	20.02×10^{-3}
Density (kg/m^3)	8960	7870	2690	5320	2330
Melting point (°C)	1083	1530	660	958.5	1412
α (μm/(mK))	16.7	12.1	23.5	5.75	7.6
Thermal conductivity at 20°C (W/mK)	410	81	238	63	156
Cohesive Energy (J/kg mol)	3.38×10^8	4.05×10^8	-	3.72×10^8	4.39×10^8
Resisivity at 20°C (Ωm)	1.72×10^{-8}	10×10^{-8}	2.69×10^{-8}	-	-
Temperature coefficient of resistance (K^{-1})	+0.0043	+0.0065	+0.0042	-	-
Ionisation Potential (eV)	7.72	7.87	5.99	-	-
Work function (eV)	4.51	4.60	4.19	-	-
Mobility (m^2/(Vs)) electrons holes	-	-	-	0.38 0.18	0.19 0.05
Energy gap (room temperature)	-	-	-	0.67	1.107
Density of states effective mass electrons holes	-	-	-	$0.35\,m_e$ $0.56\,m_e$	$0.58\,m_e$ $1.06\,m_e$

6.10 *Elements in Alphabetical Order of Symbol*

Symbol	Name	Atomic Number	Atomic Mass	Symbol	Name	Atomic Number	Atomic Mass
A or Ar	Argon	18	39.948	Mn	Manganese	25	54.9380
Ac	Actinium	89	-	Mo	Molybdenum	42	95.94
Ag	Silver	47	107.870	N	Nitrogen	7	14.0067
Al	Aluminium	13	26.9815	Na	Sodium	11	22.9898
Am	Americium	95	-	Nb	Niobium	41	92.906
As	Arsenic	33	74.9216	Nd	Neodymium	60	144.24
At	Astatine	85	-	Ne	Neon	10	20.183
Au	Gold	79	196.967	Ni	Nickel	28	58.71
B	Boron	5	10.811 ±0.003	No	Nobelium	102	-
				Np	Neptunium	93	-
Ba	Barium	56	137.34	O	Oxygen	8	15.9994 ±0.0001
Be	Beryllium	4	9.0122				
Bi	Bismuth	83	208.980	Os	Osmium	76	190.2
Bk	Berkelium	97	-	P	Phosphorus	15	30.9738
Br	Bromine	35	79.909	Pa	Protoactinium	91	-
C	Carbon	6	12.01115 ±0.00005	Pb	Lead	82	207.19
				Pd	Palladium	46	106.4
Ca	Calcium	20	40.08	Pm	Promethium	61	-
Cd	Cadmium	48	112.40	Po	Polonium	84	-
Ce	Cerium	58	140.12	Pr	Praseodymium	59	140.907
Cf	Californium	98	-	Pt	Platinum	78	195.09
Cl	Chlorine	17	35.453	Pu	Plutonium	94	-
Cm	Curium	96	-	Ra	Radium	88	-
Co	Cobalt	27	58.9332	Rb	Rubidium	37	85.47
Cr	Chromium	24	51.996	Re	Rhenium	75	186.2
Cs	Caesium	55	132.905	Rh	Rhodium	45	102.905
Cu	Copper	29	63.54	Rn	Radon	86	-
Dy	Dysprosium	66	162.50	Ru	Ruthenium	44	101.07
Er	Erbium	68	167.26	S	Sulphur	16	32.064 ±0.003
Es	Einsteinium	99	-				
Eu	Europium	63	151.96	Sb	Antimony	51	121.75
F	Fluorine	9	18.9984	Sc	Scandium	21	44.956
Fe	Iron	26	55.847	Se	Selenium	34	78.96
Fm	Fermium	100	-	Si	Silicon	14	28.086 ±0.001
Fr	Francium	87	-				
Ga	Gallium	31	69.72	Sm	Samarium	62	150.35
Gd	Gadolinium	64	157.25	Sn	Tin	50	118.69
Ge	Germanium	32	72.59	Sr	Strontium	38	87.62
H	Hydrogen	1	1.00797 ±0.00001	Ta	Tantalum	73	180.948
				Tb	Terbium	65	158.924
He	Helium	2	4.0026	Tc	Technetium	43	-
Hf	Hafnium	72	178.49	Te	Tellurium	52	127.60
Hg	Mercury	80	200.59	Th	Thorium	90	232.038
Ho	Holmium	67	164.930	Ti	Titanium	22	47.90
I	Iodine	53	126.9044	Tl	Thallium	81	204.37
In	Indium	49	114.82	Tm	Thulium	69	168.934
Ir	Iridium	77	192.2	U	Uranium	92	238.03
K	Potassium	19	39.102	V	Vanadium	23	50.942
Kr	Krypton	36	83.80	W	Tungsten	74	183.85
La	Lanthanum	57	138.91	Xe	Xenon	54	131.30
Li	Lithium	3	6.939	Y	Yttrium	39	88.905
Lu	Lutetium	71	174.97	Yb	Ytterbium	70	173.04
Md	Mendeleevium	101	-	Zn	Zinc	30	65.37
Mg	Magnesium	12	24.312	Zr	Zirconium	40	91.22

Note: the values given normally indicate the mean atomic mass of the mixture of isotopes found in nature. Particular attention is drawn to the values for hydrogen, boron, carbon, oxygen, silicon and sulphur, where the deviation shown is due to variation in relative concentration of isotopes.

6.11 *Polymer Names and Structures*

Common Name	Chemical Name	Abbre-viation	Structure	State
Polythene	Polyethylene HD	HDPE	$[\text{--}CH_2\text{--}CH_2\text{--}]_n$	Crystalline
PVC	Polyvinyl chloride	PVC	$\left[\begin{array}{c} \text{--}CH_2\text{--}CH\text{--} \\ \mid \\ Cl \end{array}\right]_n$	Amorphous/ slightly crystalline
Polystyrene	Polystyrene	PS	$\left[\begin{array}{c} \text{--}CH_2\text{--}CH\text{--} \\ \mid \\ \bigcirc \end{array}\right]_n$	Amorphous/ crystalline
Perspex	Polymethyl methacrylate	PMMA	$\left[\begin{array}{c} CH_3 \\ \mid \\ \text{--}CH_2\text{--}C\text{--} \\ \mid \\ O{=}C\text{--}O\text{-}CH_3 \end{array}\right]_n$	Amorphous
PTFE	Polytetra-fluorethylene	PTFE	$[\text{--}CF_2\text{--}CF_2\text{--}]_n$	Crystalline
Polypropylene	Polypropylene	PP	$\left[\begin{array}{c} \text{--}CH_2\text{--}CH\text{--} \\ \mid \\ CH_3 \end{array}\right]_n$	Amorphous/ crystalline
Nylon	Polyamide 6:6	PA	$\left[\text{--}NH\text{-}(CH_2)_6\text{-}NH\text{--}\underset{O}{\overset{\|}{C}}\text{-}(CH_2)_4\text{-}\underset{O}{\overset{\|}{C}}\text{--}\right]_n$	Crystalline
Rubber	Polyisoprene	PI	$\left[\begin{array}{c} CH_3 \\ \mid \\ \text{--}CH_2\text{--}C{=}CH\text{--}CH_2\text{--} \end{array}\right]_n$	Elastomer
Polycarbonate	Polycarbonate	PC	$\left[\text{--}O\text{--}\underset{O}{\overset{\|}{C}}\text{--}O\text{--}\bigcirc\text{--}\underset{CH_3}{\overset{CH_3}{\mid}}\!\!C\!\!\underset{}{\overset{}{}}\bigcirc\text{--}\right]_n$	Amorphous/ crystalline
Bakelite	Phenol formaldehyde resin	PF	$\left[\begin{array}{c} OH \qquad OH \\ \bigcirc\text{--}CH_2\text{--}\bigcirc\text{--}CH_2\text{--} \\ \mid \qquad\quad \mid \\ CH_2 \qquad CH_2 \end{array}\right]_n$	Amorphous

7. Properties of Materials

7.1 *Solids* (Unless otherwise stated, all properties are given for a temperature of 20°C)

Metals													
	ρ	E	G	K	ν	α	$\sigma_{y(0.2)}$ *	σ_f *	K_{IC}	Melting Range	c_p	k	Resistivity
	kg/m^3	GN/m^2	GN/m^2	GN/m^2		$\mu m/(mK)$	MN/m^2	MN/m^2	$MN/m^{3/2}$	°C	$kJ/(kgK)$	$W/(mK)$	Ωm
Aluminium 99.99%	2700	69	25.5	75.5	0.34	23.5	<25	<58	45	>660	0.9	244	2.7×10^{-8}
Aluminium Alloy 2024	2770	74	27.6	77.1	0.3	23	325-345	470-480	10-50	500-640	0.88	151	5.7×10^{-8}
5083	2670	71	26.8	73.9	0.3	24.5	140-285	310-380	30-50	580-645	0.89	109	6.1×10^{-8}
7075TF	2800	72	27.2	74.1	0.3	32.9	<505	<570	20-70	475-630	0.88	130	5.2×10^{-8}
Copper 99.9%	8900	125	48.3	138	0.35	17	60-325	220-385	>100	1083	0.39	393	1.7×10^{-8}
Brass 65/35	8450	102	37.3	115	0.35	21	290-300	460-480	30-100	900-920	0.38	125	6.2×10^{-8}
Cupro-Nickel 70/30	8950	147	54.4	163	0.35	16	170-200	400-430	>100	1170-1240	0.34	29	$\sim20\times10^{-8}$
Mild Steel 055M15	7860	207	82.2	169	0.29	12.4	>150	>310	140	>1550	0.46	50	16.9×10^{-8}
0.4%C Steel 080M40	7860	210	80	165	0.30	11.9	245-280	510-550	20-50	>1550	0.46	50	17.1×10^{-8}
Stainless Steel 304S15	7930	205	84	166	0.28	17.0	>195	>480	>130	>1550	0.46	15	70×10^{-8}

* the lower values of σ_y and σ_f refer to materials in the as-manufactured state, the higher values refer to materials which have been heat treated or mechanically worked.

Polymers								
	ρ	E	α	σ_f *	K_{IC}	Glass Tr. Temp. T_g	c_p	k	Resistivity
	kg/m^3	GN/m^2	$\mu m/(mK)$	MN/m^2	$MN/m^{3/2}$	°C (typical)	$kJ/(kgK)$	$W/(mK)$	Ωm
Polythene HD	970	0.55-1.0	150-300	20-37	2-5	−90	2.1	0.52	$>10^{14}$
PVC	1400	2.4-3.0	50-70	40-60	2	87		0.15	$>10^{12}$
Polystyrene	1050	3.0-3.3	70-100	35-68	2	100	1.3-1.5	0.1-0.15	$>10^{15}$
Polymethyl methacrylate 'Perspex'	1200	3.3	50-70	80-90	1.5	105	1.5	0.2	$>10^{14}$
PTFE	2200	0.35	70-100	17-28		126	1.05	0.25	$>10^{15}$
Polypropylene	910	1.2-1.7	100-300	50-70	3.5	−18	1.9	0.2	$>10^{13}$
Polycarbonate	1200	2.2-4.0	60-70	50-60	5-8	150	1.2-1.3	0.19	$>10^{16}$
Nylon 6.6	1150	2.0-3.5	100-150	60-110	3-5	50	1.9	0.2-0.25	$>10^{14}$
Polyisoprene (natural rubber)	910	0.002-0.1	~600	~10		−70	~2.5	~0.15	$>10^{10}$
Bakelite	1300	8	20-60	35-55		-	1.5-1.7	0.12-0.24	$>10^{10}$

* σ_f range is related to the exact composition of the polymer and the degree of drawing under stress.

Other Materials											
ρ	E	G	K	ν	α	σ_f	K_{IC}	Melting	c_p	k	Resistivity
kg/m^3	GN/m^2	GN/m^2	GN/m^2		μm/(mK)	MN/m^2	MN/m$^{3/2}$	°C	kJ/(kgK)	W/(mK)	Ωm
Alumina	3800	350			0.24	9	140-200	3-5	>2000	1.05	29	>10^{12}
Quartz	2650	73	31	37	0.17	0.5		~0.5	>1600	0.73	12	>10^{14}
Glass (soda)	2480	74	29	41	0.22	8.5	30-90	0.7	1000	0.99	1	>10^{14}
Concrete	2400	14			0.1-0.2	13	~20*	0.2		1.1	1.1	
Firebrick	2100					3-10		~1.0		0.81	0.4	
Ice at 0°C	917	9.5	8.8	3.9	0.33	55	9	~0.15	0	2.05	2.2	>10^5
Wood (along grain)	400-800	8-13				3-5	10-100	0.05-1.0		~2.0	~0.15	

* failure stress in compression

7.2 *Liquids*

	ρ	c_p	μ	k	σ	β	K
	kg/m^3	kJ/(kgK)	Ns/m^2	W/(mK)	N/m	K^{-1}	GN/m^2
Water	1000	4.19	1.002×10^{-3}	0.6	0.073	0.21×10^{-3}	2.2
Mercury	13 600	0.14	1.55×10^{-3}	8.7	0.51	0.18×10^{-3}	2.8
Castor oil	960	2.20	3.14×10^{-3}	0.18	0.039		2.2
Benzene	880	1.80	0.656×10^{-3}	0.16	0.029	1.34×10^{-3}	1.0
Ethyl alcohol	790	2.86	1.20×10^{-3}	0.19	0.022	1.08×10^{-3}	1.0
Engine oil (typical)	890	1.9	80×10^{-3}	0.15		0.8×10^{-3}	1.5
Freon 12	1350	0.96	0.273×10^{-3}	0.073			

7.3 *Gases*

Properties at 20°C, 1 bar

	ρ	c_p	μ	k	R	$\gamma = c_p/c_v$
	kg/m^3	kJ/(kgK)	Ns/m^2	W/(mK)	kJ/(kgK)	
hydrogen	0.082	14.3	0.88×10^{-5}	0.18	4.16	1.40
helium	0.164	5.23	1.96×10^{-5}	0.14	2.08	1.66
nitrogen	1.16	1.04	1.76×10^{-5}	0.026	0.294	1.40
oxygen	1.31	0.91	2.03×10^{-5}	0.026	0.260	1.40
carbon dioxide	1.8	0.84	1.47×10^{-5}	0.017	0.190	1.28
air	1.19	1.005	1.82×10^{-5}	0.026	0.287	1.40

Composition of air

Air (28.96)*	Nitrogen N_2 (28.013)	Oxygen O_2 (31.999)	Argon Ar (39.948)	Carbon dioxide CO_2 (44.010)
By volume	0.7809	0.2095	0.0093	0.0003
By mass	0.7553	0.2314	0.0128	0.0005

* the numbers in brackets are the molecular masses of the gases.

Critical Constants

	Molecular mass	T_c (K)	P_c (bar)	ρ_c (kg/m³)
hydrogen	2.02	33.3	13.0	31
helium (4)	4.00	5.3	2.29	69.3
water vapour	18.02	647.30	221.2	318.3
nitrogen	28.01	126.1	33.9	311
oxygen	32.00	154.4	50.4	430
carbon dioxide	44.01	304.15	73.8	468

7.4 *Fuels*

Gases

	Composition by volume %							Relative density (air = 1)	Calorific value at 1.01325 bar, 15°C MJ/m³		Theoretical air
	N_2	H_2	CH_4	C_2H_6	C_3H_8	C_4H_{10}	C_3H_6		Gross	Net	vol/- vol
hydrogen		100						0.0696	12.10	10.22	2.38
methane			100					0.5537	37.71	33.95	9.52
North Sea gas	1.5		94.4	3.0	0.5	0.2		0.589	38.62	34.82	9.75
propane*				1.5	91.0	2.5	5.0	1.523	98.87	86.43	23.76
butane*			0.1	0.5	7.2	87.0	4.2	1.941	117.75	108.69	29.92

* commercial liquid petroleum gas (LPG)

Liquids

	Composition by mass %			Density at 15°C kg/m³	Calorific value at 15°C MJ/kg	
	C	H	S		Gross	Net
propane*	82.0	18.0		505	50.0	46.3
butane*	81.9	17.0		575	49.3	45.8
petrol (gasoline)	85.5	14.4	0.1	733	46.9	43.7
kerosene (parafin)	85.9	14.0	0.1	780	46.5	43.4
diesel (gas oil)	85.7	13.4	0.9	840	45.4	42.4

* commercial liquid petroleum gas (LPG)

8. Earth and the Environment
8.1 *Geophysical Constants*

Mean Depth (km)		Mass (kg)	
Earth Centre	6371	Whole earth	5.9736×10^{24}
Solid core	5150	Inner core	9.84×10^{22}
Core-mantle boundary	2891	Outer core	1.841×10^{24}
Upper mantle	660	Mantle	4.002×10^{24}
Oceanic Lithosphere	80	Lower Mantle	2.940×10^{24}
Oceanic crust	7.1	Upper Mantle	1.062×10^{24}
Continental crust	39	Continental Crust	2.97×10^{22}
Ocean	3.77	Oceans	1.40×10^{21}
Continental elevation	0.825	Polar Ice and Glaciers	4.34×10^{19}
Surface area (m^2)		Atmosphere	5.1×10^{18}
Whole earth	5.10×10^{14}	**Mean density (kg m^{-3})**	
Land	1.48×10^{14}	Whole earth	5526
Fluxes		Core	10 957
Oceanic crust at ridge crests ($km^2 yr^{-1}$) 3.5		Lower mantle	4903
Rivers and runoff (kg yr^{-1})	3.6×10^{16}	Upper mantle	3604
Mechanical erosion (kg yr^{-1})	2.0×10^{13}		

8.2 *The Geological Time Scale*

Eon	Era	Period	Epoch	End – Start (Ma)	Relative Climate	UK Name
Phanerozoic	Cenozoic	Quaternary	Recent/ Holocene	0 – 0.018	Warm (current)	Flandrian (Holocene)
			Pleistocene	0.018 – 1.6	Cold	Devensian
		Tertiary	Pliocene	1.6 – 5.1	Warm	Ipswichian
			Miocene	5.1 – 24	Cold	Wolstonian
			Oligocene	24 – 38	Warm	Hoxnian
			Eocene	38 – 55	Cold	Anglian
			Paleocene	55 – 65	Warm	Cromerian
	Mesozoic	Cretaceous		65 – 144	Cold	Beestonian
		Jurassic		144 – 213	Warm	Pastonian
		Triassic		213 – 248	Cold	Baventian
	Paleozoic	Permian		248 – 286	Warm	Antian
		Carboniferous		286 – 360	Cold	Thurnian
		Devonian		360 – 408		Ludhamian
		Silurian		408 – 438		Waltonian
		Ordovician		438 – 505		
		Cambrian		505 – 570		
Proterozoic		(together comprising Precambrian)		570 – 2700		
Archean				2700 – 4000		
Hadean				4000 – 4560		

Note: the right-hand columns are the **Subdivisions of Quaternary** (Relative Climate / UK Name).

8.3 *Average Composition of Bulk Earth, Continental Crust, Oceans and Rivers*

Element	Atomic Number	Atomic Mass	Earth (ppm)	Continental Crust (ppm)	Ocean (µg/l)	Rivers (µg/l)
H	1	1.01	67.5		1.11×10^{8}	
He	2	4.00			7.20×10^{-3}	
Li	3	6.94	1.08	11	178	3.0
Be	4	9.01	0.046		2.25×10^{-4}	0.001
B	5	10.81	0.203		4390	
C	6	12.01	309		2640	
N	7	14.01	27		8540	
O	8	16.00	305 000	450 000	2400	
F	9	19.00	10.1		1270	1.0
Ne	10	20.18			3.96×10^{-4}	
Na	11	22.99	1800	23 700	1.08×10^{7}	6000
Mg	12	24.31	153 900	26 500	1.26×10^{6}	4100
Al	13	26.98	15 900	83 600	0.162	50
Si	14	28.09	171 000	276 100	2810	6500
P	15	30.97	1101	873	60.0	20
S	16	32.06	6344		8.63×10^{5}	3700
Cl	17	35.45	11.5		1.89×10^{7}	7800
Ar	18	39.95			599	
K	19	39.10	162	15 772	3.89×10^{5}	3000
Ca	20	40.08	17 100	45 700	4.14×10^{5}	20 000
Sc	21	44.96	10.93	22	6.00×10^{-4}	4×10^{-3}
Ti	22	47.90	813	4197	4.79×10^{-3}	3
V	23	50.94	94.4	131	1.78	0.9
Cr	24	52.00	3720	119	0.208	1
Mn	25	54.94	1680	852	0.192	7
Fe	26	55.85	303 000	51 300	5.59×10^{-2}	40
Co	27	58.93	838	25	2×10^{-3}	0.1
Ni	28	58.71	17 600	51	0.480	0.3
Cu	29	63.54	60.9	24	0.2	7
Zn	30	65.37	37.1	73	0.379	20
Ga	31	69.72	2.7	16	6.97×10^{-4}	9×10^{-2}
Ge	32	72.59	7.24		5.08×10^{-2}	5×10^{-3}
As	33	74.92	1.669		1.72	2
Se	34	78.96	2.65		0.118	6×10^{-2}
Br	35	79.91	0.034		6.70×10^{4}	20
Kr	36	83.80			0.293	
Rb	37	85.47	0.405	58	124	1
Sr	38	87.62	13.4	325	8000	70
Y	39	88.91	2.90	20	8.66×10^{-3}	4×10^{-2}
Zr	40	91.22	7.07	123	1.20×10^{-2}	
Nb	41	92.91	0.444	12	4.65×10^{-3}	
Mo	42	95.94	1.660		11.0	0.6
Ru	44	101.07	1.310			

Element	Atomic Number	Atomic Mass	Earth (ppm)	Continental Crust (ppm)	Ocean (μg/l)	Rivers (μg/l)
Rh	45	102.91	0.242			
Pd	46	106.42	1.015		4.26×10^{-5}	
Ag	47	107.87	0.0544		2.37×10^{-2}	0.3
Cd	48	112.40	0.076		6.74×10^{-2}	1×10^{-2}
In	49	114.82	0.0068		1.15×10^{-4}	
Sn	50	118.69	0.2511		4.75×10^{-4}	4×10^{-2}
Sb	51	121.75	0.0462		0.122	7×10^{-2}
Te	52	127.60	0.2857		1.02×10^{-4}	
I	53	126.90	6.8		57.1	7
Xe	54	131.30			6.56×10^{-2}	
Cs	55	132.91	0.014	2.6	0.300	2×10^{-2}
Ba	56	137.34	4.455	390	11.7	20
La	57	138.91	0.437	18	3.88×10^{-3}	5×10^{-2}
Ce	58	140.12	1.131	42	7.01×10^{-4}	8×10^{-2}
Pr	59	140.91	0.171	5	7.72×10^{-4}	7×10^{-3}
Nd	60	144.24	0.844	20	2.6×10^{-3}	2×10^{-3}
Sm	62	150.35	0.274	3.9	5.1×10^{-4}	8×10^{-3}
Eu	63	151.96	0.104	1.2	1.34×10^{-4}	2×10^{-3}
Gd	64	157.25	0.367	3.6	7.94×10^{-4}	9×10^{-3}
Tb	65	158.92	0.067	0.056	1.81×10^{-4}	1×10^{-3}
Dy	66	162.50	0.455	3.5	9.47×10^{-4}	7×10^{-3}
Ho	67	164.93	0.101	0.76	2.79×10^{-4}	1×10^{-3}
Er	68	167.26	0.296	2.2	9.17×10^{-4}	4×10^{-3}
Tm	69	168.93	0.046	0.32	1.63×10^{-4}	6×10^{-4}
Yb	70	173.04	0.297	2	9.81×10^{-4}	4×10^{-3}
Lu	71	174.97	0.046	0.33	1.80×10^{-4}	6×10^{-4}
Hf	72	178.49	0.191	3.7	1.61×10^{-4}	
Ta	73	180.95	0.025	1.1	2.53×10^{-3}	
W	74	183.85	0.173		0.110	3×10^{-2}
Re	75	186.21	75.3		7.2×10^{-3}	
Os	76	190.23	0.9		1.70×10^{-6}	
Ir	77	192.20	0.835		1.15×10^{-6}	
Pt	78	195.09	1.886		1.17×10^{-4}	
Au	79	196.97	0.164		9.85×10^{-6}	2×10^{-3}
Hg	80	200.59	0.0231		1.40×10^{-3}	7×10^{-2}
Tl	81	204.37	0.0122		1.23×10^{-2}	
Pb	82	207.19	0.232	12.6	5.00×10^{-4}	1
Bi	83	208.98	9.853		1.04×10^{-5}	
Th	90	232.04	0.054	5.6	5.00×10^{-5}	
U	92	238.03	0.014	1.42	3.20	4×10^{-2}

Note: the names of the elements, in alphabetical order, can be found in section 6.10. The periodic table of the elements may be found in section 6.9.

8.4 *Rocks*

Classification of igneous rocks

Fine Grained: Extrusive or Volcanic	Rhyolite	Dacite	Andesite	Basalt	
Coarse Grained: Intrusive or Plutonic	Granite	Granodiorite	Diorite	Gabbro	Peridotite

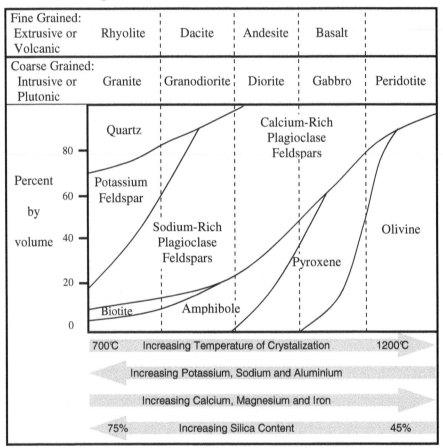

Metamorphic rocks

Parent	Product
mudstone	slate → schist → gneiss
sandstone	(meta)quartzite
limestone	marble
basalt	amphibolite
granite	gneiss

Composition of the main Sedimentary Rocks

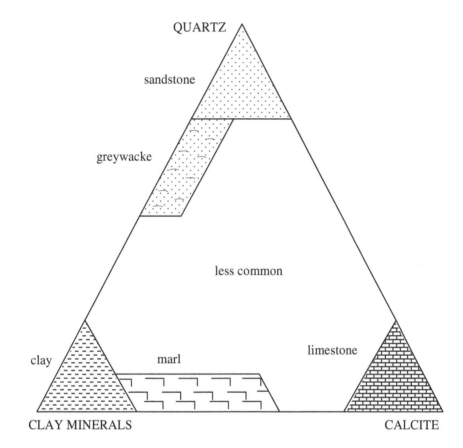

Grain size of Clastic Sedimentary Rocks (see also soil classification, section 12.1)

Particle Size (mm)	Sediment	Rock
Coarse	Gravel	
> 256	Boulder	
64 – 256	Cobble	} Conglomerate
2 – 64	Pebble	
Medium: 0.062 – 2	Sand	Sandstone
Fine	Mud	
0.0039 – 0.062	Silt	Siltstone
< 0.0039	Clay	{ Mudstone (blocky fracture) Shale (breaks along bedding) Claystone

8.5 *Isotopes*

Radioactive Decay

Radioactive decay equation: $D = D_0 + N(e^{\lambda t} - 1)$

D_0 number of daughter nuclides at time zero
D number of daughter nuclides produced after time t
N number of parent nuclides remaining after time t
λ decay constant of the parent nuclide

Half-life of parent nuclide: $t_{1/2} = \dfrac{\ln(2)}{\lambda} = \dfrac{0.693}{\lambda}$

Decay constants and half-lives for a selection of environmentally useful radioactive isotopes

Parent nuclide	Decay constant (years^{-1})	Half-life (years)	Daughter nuclide
^{14}C	1.21×10^{-4}	5730	14N
^{40}K	5.54×10^{-10}	1.25×10^{9}	^{40}Ar, ^{40}Ca
^{138}La	2.67×10^{-12}	2.59×10^{11}	^{138}Ce, ^{138}Ba
^{176}Lu	1.94×10^{-11}	3.6×10^{10}	^{176}Hf
^{190}Os	1.54×10^{-12}	4.50×10^{11}	^{186}Os
^{231}Pa	212×10^{-5}	32 760	^{227}Ac
^{210}Pb	3.11×10^{-2}	22.3	^{206}Pb, ^{210}Bi
^{226}Ra	4.33×10^{-4}	1600	^{222}Rn
^{87}Rb	1.42×10^{-11}	4.8×10^{10}	^{87}Sr
^{187}Re	1.64×10^{-11}	4.23×10^{10}	^{187}Os
^{147}Sm	6.54×10^{-12}	1.06×10^{11}	^{143}Nd
^{230}Th	9.20×10^{-5}	75 300	^{226}Ra
^{232}Th	4.95×10^{-11}	1.4×10^{10}	^{208}Pb, ^{4}He
^{234}Th	10.5	0.066	^{234}Pa
^{234}U	2.82×10^{-6}	2.46×10^{5}	^{230}Th
^{235}U	9.85×10^{-10}	7.07×10^{8}	^{207}Pb, ^{4}He
^{238}U	1.55×10^{-10}	4.47×10^{9}	^{206}Pb, ^{4}He

Stable Isotopes
δ-notation

δ = permil variation in stable isotope ratio R from some standard

$$\delta = \left[\frac{R_{sample} - R_{standard}}{R_{standard}} \right] \times 1000\%o$$

Fractionation factor

The fractionation of isotopes between two phases A and B is described by the fractionation factor $\alpha_{A-B} = \dfrac{R_A}{R_B}$ or by $\Delta_{A-B} \approx 10^3 \ln \alpha_{A-B}$

Notation for isotopic ratio and standards for common stable isotope systems

Element	Notation	Ratio	Standard
Hydrogen	δD	D/H (^{2}H/^{1}H)	Standard Mean Ocean Water (SMOW)
Helium	δ^3He	^{3}He/^{4}He	Air
Lithium	δ^6Li	^{6}Li/^{7}Li	International standard of synthetic lithium carbonate (NBS L-SVEC)
Boron	δ^{11}B	^{11}B/^{10}B	International standard 951 (NBS 951)
Carbon	δ^{13}C	^{13}C/^{12}C	Pee Dee Belemnite (PDB)
Nitrogen	δ^{15}N	^{15}N/^{14}N	Atmosphere
Oxygen	δ^{18}O	^{18}O/^{16}O	SMOW, PDB ($\delta^{18}O_{PDB} = 1.03086\,\delta^{18}O_{SMOW} + 30.86$)
	δ^{17}O	^{17}O/^{16}O	SMOW
Chlorine	δ^{37}Cl	^{37}Cl/^{35}Cl	Seawater
Sulphur	δ^{34}S	^{34}S/^{32}S	Canyon Diablo Triolite (CDT)

8.6 *The Ocean*

Potential Temperature

Potential temperature θ : the temperature a parcel of seawater would have if moved adiabatically to the surface.

Density

Sigma notation:

$$\sigma = \rho_{S,\,\theta,\,p} - 1000$$

where ρ is the density in kg m^{-3}, considered as a function of salinity S (ppt), potential temperature θ (°C) and pressure p (decibars). (Note: 1 decibar is approximately equivalent to 1 m depth in sea water.)

Sigma-t	$\sigma_t = \rho_{S,\,\theta,\,0} - 1000$	Upper ocean, above 1 km
Sigma-2	$\sigma_2 = \rho_{S,\,\theta,\,2000} - 1000$	Depths from 1 to 3 km
Sigma-4	$\sigma_4 = \rho_{S,\,\theta,\,4000} - 1000$	Below 3 km

Specific volume $\alpha = \dfrac{1}{\rho}$; *Specific volume anomaly* $\delta = \alpha_{S,\,T,\,p} - \alpha_{35,\,0,\,p}$

Thermosteric anomaly $\Delta = \alpha_{S,\,T,\,p} - \alpha_{35,\,0,\,0}$ is equivalent to Sigma-t (above).

Depth and Hydrostatic Pressure   See under Fluid Mechanics, section 9.4

Dynamic or Geopotential Height

Dynamic height $D = \int_{p_1}^{p_2} \alpha dp$ J kg^{-1} represents the energy of the water column per unit mass.

Speed of sound in sea water (salinity $S = 35$ ppt) = 1507 ms^{-1}

8.7 *The Atmosphere*

Composition of Air – see section 7.3

Environmental effects of Refrigerants and other Gases

Name	Chemical Formula	Boiling Point (at atmospheric pressure) (°C)	Ozone Depletion Potential (Montreal controlled substances)	Greenhouse Warming Potential (IPCC 100 year horizon)
Carbon Dioxide	CO_2	−78	*	1
Dichlorodifluoromethane, CFC12 "Freon"	CCl_2F_2	−29.8	1	10600
Chlorodifluoromethane, HCFC22	$CHClF_2$,	−40.8	0.055	1700
Tetrafluoroethane, HFC134a	$H_2FC\text{-}CF_3$	−26.6	*	1300
Isobutane, R600a	$(CH_3)_2CH\text{-}CH_3$	−11.6	*	< 4
Propane, R290	C_2H_6	−42.1	*	< 4
Methane	CH_4	−161.6	–	23
Halon 1211	CF_2ClBr	−4	3	1300
Ammonia	NH_3	−33.5	*	0
Carbon Tetrachloride	CCl_4	77	1.1	1800
1,1,1-trichloroethane (methyl chloroform)	CH_3CCl_3	74	0.1	140
Nitrous oxide	N_2O	−88.5	–	296
Sulphur hexafluoride	SF_6	−63.9 (sublimation)	–	22200
Water (vapour)	H_2O	100	–	Hitherto considered a natural variable – may need to be reconsidered

* These refrigerants are not Montreal controlled substances and are considered to have zero Ozone Depletion Potential

9. Thermodynamics and Fluid Mechanics

9.1 *Temperatures at the primary fixed points*

Normal boiling point of oxygen (oxygen point)	−182.97°C
Triple point of water	0.01°C
Normal boiling point of water (steam point)	100.00°C
Normal boiling point of sulphur (sulphur point)	444.60°C
Normal melting point of silver (silver point)	960.80°C
Normal melting point of gold (gold point)	1063.00°C

9.2 *Thermodynamic Relationships*

In all thermodynamic relationships, temperatures and pressures are absolute. Symbols for properties in capitals refer to a given mass of material, those in lower case are per unit mass ("specific").

First Law:	$\mathrm{d}Q - \mathrm{d}W = \mathrm{d}U$
Work:	$\mathrm{d}W = p\,\mathrm{d}V$
Enthalpy:	$H = U + pV$ or $h = u + pv$
For a reversible process,	$\mathrm{d}S = \left(\dfrac{\mathrm{d}Q}{T}\right)_{rev}$ or $\mathrm{d}Q = T\mathrm{d}S$
For a homogeneous fluid	$T\mathrm{d}s = \mathrm{d}u + p\,\mathrm{d}v = \mathrm{d}h - v\,\mathrm{d}p$
Specific heat at constant volume:	$c_v = \left(\dfrac{\mathrm{d}u}{\mathrm{d}T}\right)_v$
Specific heat at constant pressure:	$c_p = \left(\dfrac{\mathrm{d}h}{\mathrm{d}T}\right)_p$
Ratio of specific heats:	$\gamma = \dfrac{c_p}{c_v}$ (for values, see table 7.3)
Helmholtz function:	$F = U - TS$ or $f = u - Ts$
Gibbs function, Gibbs free energy:	$G = H - TS$ or $g = h - Ts$

Perfect Gas Relationships	
$pv = RT$ or $p = \rho RT$	$PV = mRT$
$pv_0 = R_0 T$	$MR = R_0$
$\Delta U = mc_v(T_2 - T_1)$	$\Delta H = mc_p(T_2 - T_1)$
$\Delta S = mc_v \ln\left(\dfrac{T_2}{T_1}\right) + mR \ln\left(\dfrac{v_2}{v_1}\right)$	$\Delta S = mc_p \ln\left(\dfrac{T_2}{T_1}\right) - mR \ln\left(\dfrac{P_2}{P_1}\right)$
$c_p - c_v = R$ (for values see table 7.3)	$\dfrac{\gamma - 1}{\gamma} = \dfrac{R}{c_p}$

Continuity equation: $\qquad\qquad\qquad \dot{m} = \rho A c$

Steady Flow Energy equation:

$$\frac{\dot{Q} - \dot{W}}{\dot{m}} = h_2 - h_1 + \tfrac{1}{2}(c_2^2 - c_1^2) + g(z_2 - z_1)$$

Availability (closed system):

$$(A_1 - A_0) = (U_1 + p_0 V_1 - T_0 S_1) - (U_0 + p_0 V_0 - T_0 S_0)$$

Availability (flow process):

$$(B_1 - B_0) = (H_1 - T_0 S_1) - (H_0 - T_0 S_0)$$

Irreversibility: $\qquad\qquad\qquad I = W_{rev} - W$

Van der Waal's equation: $\qquad (p + \dfrac{a}{v^2})(v - b) = RT$

Entropy: $\qquad\qquad\qquad S = k \ln p, \ \text{where } k = R_0/N$

Reversible (Carnot) engine efficiency: $1 - \dfrac{T_{sink}}{T_{source}}$

Engine indicated power: $\qquad\qquad P_i = p_m v_s N_c$

Maximum work of a reaction:

$$W_{max} = G_{react} - G_{prod} = R_0 T \ \ln(K_p p^n)$$

For reversible polytropic process in a closed system:

$$p V^n = \text{const}$$

$$W = \frac{(p_1 V_1 - p_2 V_2)}{(n - 1)} \qquad (n \neq 1)$$

Additionally, if a perfect gas,

$$W = \frac{m R (T_1 - T_2)}{(n - 1)} \qquad \frac{T_2}{T_1} = \left(\frac{p_2}{p_1}\right)^{\frac{n-1}{n}} = \left(\frac{V_1}{V_2}\right)^{n-1} \qquad Q = \frac{\gamma - n}{\gamma - 1} W$$

Adiabatic reversible (isentropic reversible): $n = \gamma$

Isothermal reversible: $\qquad n = 1 \qquad W = Q = pV \ln\left(\dfrac{V_2}{V_1}\right)$

Maxwell Relations

$$\left(\frac{\partial T}{\partial v}\right)_s = -\left(\frac{\partial p}{\partial s}\right)_v \qquad \left(\frac{\partial T}{\partial p}\right)_s = \left(\frac{\partial v}{\partial s}\right)_p$$

$$\left(\frac{\partial p}{\partial T}\right)_v = \left(\frac{\partial s}{\partial v}\right)_T \qquad \left(\frac{\partial v}{\partial T}\right)_p = -\left(\frac{\partial s}{\partial p}\right)_T$$

Wet Vapours	Air-vapour Mixtures
$h = h_f + xh_{fg}$ $u = u_f + xu_{fg}$ $s = s_f + xs_{fg}$	Relative humidity (percentage saturation): $\phi = \dfrac{p_s}{p_g}$ Specific humidty: $w = \dfrac{m_s}{m_a} = 0.622\,\dfrac{p_s}{p - p_s}$

9.3 *Heat Transfer*

Log mean temperature difference:

$$\Delta T_m = \frac{\Delta T_{in} - \Delta T_{out}}{\ln\left(\dfrac{\Delta T_{in}}{\Delta T_{out}}\right)}$$

Heat transfer coefficient:

$$h = \frac{\dot{Q}}{A\Delta T}$$

Emissivity:

$$\varepsilon = \frac{q}{q_b}$$

Conduction:

(one-dimensional):

$$\frac{\dot{Q}}{A} = -k\frac{dT}{dx}$$

(radial):

$$\frac{\dot{Q}}{l} = \frac{2\pi\,k\Delta T}{\ln\left(\dfrac{r_2}{r_1}\right)}$$

Forced convection: $Nu = f(Re,\ Pr)$

Turbulent forced convection in a tube: $Nu = 0.023\,Re^{0.8}Pr^{0.4}$

Free convection: $Nu = f(Gr,\ Pr)$

Laminar free convection on a vertical plate: $Nu = 0.59\,(Gr{\times}Pr)^{0.25}$

Radiation:

Wien's Law $\lambda_{peak}\,T = \text{const} = 0.079\ \text{mK}$

Stefan-Boltzmann Law: $q_b = \sigma T^4$ (for σ see section 2)

Grey body to black or large enclosure: $\dfrac{\dot{Q}}{A} = \sigma\varepsilon_1(T_1^4 - T_2^4)$

Large parallel grey surfaces: $\dfrac{\dot{Q}}{A} = \dfrac{\sigma(T_1^4 - T_2^4)}{\left(\dfrac{1}{\varepsilon_1} + \dfrac{1}{\varepsilon_2} - 1\right)}$

9.4 *Fluid Mechanics*

In fluid mechanics pressures can generally be either absolute or gauge. Either may be used except where specifically noted.

Statics

$$\frac{dp}{dz} = -\rho g$$

$$p = \rho g h$$

Normal force $F = \rho g \bar{h} A$

$$\varepsilon = \frac{A k^2_{\text{centroid}}}{A \bar{y}}$$

$$\overline{GM}_{\text{roll}} = \frac{A k^2}{V} - \overline{BG}$$

(k^2 about axis of roll)

F: force

$\bar{y}$

ε

$\bar{h}$

lamina of area A

G: centroid

P: centre of pressure

water line area A

M metacentre

G centre of gravity

B centre of buoyancy

immersed volume V

Dynamics

Newtonian viscosity:	$\tau = \mu \dfrac{du}{dy}$
Viscous (laminar) pipe flow:	$Q = \dfrac{\pi r^4}{8\mu}\dfrac{dp}{dx},\ f = \dfrac{16}{Re}$ (see note on page 9-10)
Stokes' law:	$D = 3\pi\mu u d$
Euler's equation:	$\dfrac{1}{\rho}\dfrac{dp}{ds} + u\dfrac{du}{ds} + g\dfrac{dz}{ds} = 0$
Bernoulli's equation:	$p + \frac{1}{2}\rho u^2 + \rho g h = \text{const} = P$
or	$\dfrac{p}{\rho g} + \dfrac{u^2}{2g} + z = \text{const} = Z$
Acceleration along a streamline	$a_s = u_s \dfrac{\partial u_s}{\partial s} + \dfrac{\partial u_s}{\partial t}$
Acceleration normal to a streamline	$a_n = \dfrac{u_s^2}{r} + \dfrac{\partial u_n}{\partial t}$
Turbulent pipe flow:	$\Delta p = 4f\dfrac{l}{d}\frac{1}{2}\rho u^2$ (see note on page 9-10)
Vortex flow:	$\dfrac{dp}{dr} = \dfrac{\rho u^2}{r}$
Free vortex: $ur = \text{const};$	Forced vortex: $\dfrac{u}{r} = \omega = \text{const}$

Constant area flow with friction (Fanno):

$$\frac{dp}{\rho} + udu + 2\frac{fu^2}{d}dl = 0$$

Total shear stress in turbulent flow:

$$\bar{\tau} = (\mu + \varepsilon)\frac{d\bar{u}}{dy}$$

Tribology

Reynolds equations for laminar flow in bearings:

$$\frac{\partial}{\partial x}\left(\frac{\rho h^3}{12\mu}\frac{\partial p}{\partial x}\right) + \frac{\partial}{\partial y}\left(\frac{\rho h^3}{12\mu}\frac{\partial p}{\partial y}\right) = \frac{1}{2}\frac{\partial}{\partial x}(\rho uh) + \frac{\partial}{\partial t}(\rho h) + \rho w$$

$$\frac{1}{r}\frac{\partial}{\partial r}\left(\frac{\rho rh^3}{12\mu}\frac{\partial p}{\partial r}\right) + \frac{1}{r^2}\frac{\partial}{\partial \phi}\left(\frac{\rho h^3}{12\mu}\frac{\partial p}{\partial \phi}\right) = \frac{1}{2}\frac{\partial}{\partial \phi}(\rho h\omega) + \frac{\partial}{\partial t}(\rho h) + \rho w$$

Petrov's Law:

$$T = \frac{\pi h\mu\omega D^3}{4c}$$

9.5 *Boundary Layers:*

Boundary Layer Thickness

Nominal	Displacement	Momentum
δ_{99} = value of y at which $u/u_m = 0.99$	$\delta^* = \int_0^\infty (1 - \frac{u}{u_m})dy$	$\theta = \int_0^\infty \frac{u}{u_m}(1 - \frac{u}{u_m})dy$

Form Factor:

$$H = \frac{\delta^*}{\theta}$$

Momentum Integral Equation

$$\tau_0 = \frac{d}{dx}(\rho u_m^2 \theta) + \rho \delta^* u_m \frac{du_m}{dx}$$

If incompressible and zero pressure gradient:

$$c_f = \frac{\tau_0}{\frac{1}{2}\rho u_m^2} = 2\frac{d\theta}{dx}; \quad C_D = \frac{2\theta_L}{L}$$

Boundary Layer Transition

Criterion - typical:

$$Re_x = 5 \times 10^5; \quad Re_\delta = 3500; \quad Re_\theta = 500$$

Turbulent Boundary Layer (zero pressure gradient):

"Friction velocity":
$$u_\tau = \sqrt{\frac{\tau_0}{\rho}}$$

In the laminar sub-layer:
$$\frac{u}{u_\tau} = \frac{y u_\tau}{\nu}$$

The edge of laminar sub-layer is at
$$\frac{u}{u_\tau} = \frac{y u_\tau}{\nu} = 11.6 \text{ (approx)}$$

Inner Law (Law of the Wall):
$$\frac{u}{u_\tau} = 5.75 \log_{10} \frac{y u_\tau}{\nu} + 5.5$$

Power law approximation:
$$\frac{u}{u_\tau} = K \left(\frac{y u_\tau}{\nu} \right)^{\frac{1}{n}} \text{ or } \frac{u}{u_m} = \left(\frac{y}{\delta} \right)^{\frac{1}{n}}$$

Typically $n = 7$, $K = 8.74$, when

$$\frac{\delta^*}{\delta} = \frac{1}{(1+n)} = 0.125 \qquad \frac{\theta}{\delta} = \frac{n}{(1+n)(2+n)} = 0.097$$

$$H = \frac{\delta^*}{\theta} = 1 + \frac{2}{n} = 1.29$$

Blasius' empirical friction formula for smooth surfaces:
$$c_f = 0.045 \, Re_x^{-0.25}$$

(note: this should be distinguished from the similar formula for smooth pipes:
$$f = c_f = 0.079 \, Re_d^{-0.25} \text{ (see also note on page 9-10))}$$

Boundary layer growth:

For zero pressure gradient:

Laminar: Blasius' exact solution	Turbulent: $\frac{1}{7}$ power law profile and $\frac{1}{4}$ power law friction	
$\dfrac{\delta_{99}}{x} = 4.91 \, Re_x^{-0.5}$	$\dfrac{\delta}{x}$	$= 0.370 \, Re_x^{-0.2}$
$\dfrac{\delta^*}{x} = 1.721 \, Re_x^{-0.5}$	$\dfrac{\delta^*}{x}$	$= 0.046 \, Re_x^{-0.2}$
$\dfrac{\theta}{x} = c_f = 0.664 \, Re_x^{-0.5}$	$\dfrac{\theta}{x}$	$= 0.036 \, Re_x^{-0.2}$
	c_f	$= 0.058 \, Re_x^{-0.2}$
$C_D = 1.328 \, Re^{-0.5}$	C_D	$= 0.072 \, Re^{-0.2}$
$H = \dfrac{\delta^*}{\theta} = 2.59$	$H = \dfrac{\delta^*}{\theta}$	$= 1.29$

9.6 *Hydraulic machines:*

Dimensionless coefficients:

Head: $K_H = \dfrac{gH}{\omega^2 D^2}$ Flow: $K_Q = \dfrac{Q}{\omega D^3}$ Power: $K_P = \dfrac{\text{Power}}{\rho \omega^3 D^5}$

Dimensionless specific speed (rad/s):	$\omega_s = \left\| \dfrac{\omega Q^{\frac{1}{2}}}{(gH)^{\frac{3}{4}}} = \dfrac{K_Q^{\frac{1}{2}}}{K_H^{\frac{3}{4}}} = \dfrac{\omega(\text{power})^{\frac{1}{2}}}{\rho^{\frac{1}{2}}(gH)^{\frac{5}{4}}} \right\|_{\eta_{max}}$
Dimensionless diameter:	$\Delta = \dfrac{D(gH)^{\frac{1}{4}}}{Q^{\frac{1}{2}}}$
NPSE:	$\dfrac{p_1 - p_v}{\rho}$
Dimensionless suction specific speed:	$K_s = \dfrac{\omega Q^{\frac{1}{2}}}{(\text{NPSE})^{\frac{3}{4}}}$
Cavitation number:	$\sigma \text{ or } k = \dfrac{p_\infty - p_v}{\frac{1}{2}\rho v_\infty^2}$
Thoma cavitation number:	$\sigma_{th} = \dfrac{p_1 - p_v}{p_2 - p_1} = \left(\dfrac{\omega_s}{K_s}\right)^{\frac{4}{3}}$

Suffices in these definitions:
∞ = reference,
v = vapour,
1 = low pressure side,
2 = high pressure side.
(Pressures must be absolute)

Cordier correlation for hydraulic machines

9.7 *Open Channel Flows:*

Hydraulic mean depth
= hydraulic radius:

$$m = R = \frac{\text{cross-sectional area}}{\text{wetted perimeter}}$$

Equivalent diameter:

$$d' = \frac{4 \times \text{cross-sectional area}}{\text{wetted perimeter}} = 4m$$

Chézy equation:

$$u = C\sqrt{RS}$$

Manning equation:

$$u = \frac{1}{n} R^{\frac{2}{3}} S^{\frac{1}{2}}$$

Steady gradually varied flow:

$$\frac{\mathrm{d}d}{\mathrm{d}x} = \frac{S_0 - S_f}{1 - \alpha \dfrac{u^2}{gd}} \quad \text{(rectangular channel)}$$

Unsteady gradually varied flow:

$$\frac{\partial d}{\partial x} + \frac{u}{g}\frac{\partial u}{\partial x} + \frac{1}{g}\frac{\partial u}{\partial t} = S_0 - S_f$$

Continuity:

$$A\frac{\partial u}{\partial x} + u\frac{\partial A}{\partial x} + T\frac{\partial d}{\partial t} = 0$$

$\left.\begin{array}{l}\\ \\ \\ \\ \\ \end{array}\right\}$ no local inflow or outflow

Conjugate depths in hydraulic jump:

$$\frac{d_2}{d_1} = \frac{1}{2}\left(\sqrt{1 + 8Fr_1^2} - 1\right) \quad \text{(rectangular channel)}$$

9.8 *High speed gas flow*

Nozzles:

Mass flow rate:

$$\dot{m} = AC_d\sqrt{\frac{2n}{(n+1)} p_0 \rho_0 \left[\left(\frac{p}{p_0}\right)^{\frac{2}{n}} - \left(\frac{p}{p_0}\right)^{\frac{n+1}{n}}\right]}$$

Critical pressure ratio:

$$\frac{p^*}{p_0} = \left(\frac{2}{n+1}\right)^{\frac{n}{n-1}}$$

Sonic velocity:

$$a = \sqrt{n\frac{p}{\rho}}$$

where $n \approx 1.3$ for steam, initially superheated
$n \approx 1.135$ for steam, initially wet or dry saturated
$n = \gamma = 1.4$ for air or diatomic gases

Perfect Gas:

Sonic Velocity:

$$\sqrt{\gamma RT}$$

Stagnation temperature:

$$T_0 = T\left[1 + \frac{(\gamma - 1)}{2}(Ma)^2\right]$$

Air ($\gamma = 1.4$) in isentropic flow:

$$\frac{\dot{m}\sqrt{T_0}}{A^* p_0} = 0.0404\,\text{kg K}^{\frac{1}{2}}/(\text{Ns})$$

9.9 *Dimensionless Groups*

Pressure coefficient:
$$C_p = \frac{p}{\frac{1}{2}\rho u^2} \text{ or } \frac{\Delta p}{\frac{1}{2}\rho u_\infty^2}$$

Force coefficient (lift, drag etc):
$$C_F = \frac{F}{\frac{1}{2}\rho u^2 A}, \quad F = L, D, \ldots$$

Discharge coefficient (venturi-meter etc):
$$C_d = \frac{Q_{\text{actual}}}{Q_{\text{ideal}}}, \quad Q_{\text{ideal}} = A_{\text{throat}}\sqrt{\frac{2\Delta p/\rho}{1-m^2}}$$
$$\text{where } m = A_{\text{throat}}/A_{\text{pipe}}$$

Loss coefficient (internal flow)
$$k = \frac{\Delta p}{\frac{1}{2}\rho u^2}$$

Pipeflow friction factor (round pipe):
(see note on page 9-10)
(for non-circular duct or free surface flow, replace d with d' or $4m$)
$\equiv$ Wall shear stress coefficient:
$$f = \frac{\Delta p}{\frac{1}{2}\rho u^2} \bigg/ \left(\frac{4l}{d}\right)$$
$$f = c_f = \frac{\tau_0}{\frac{1}{2}\rho u^2}$$

Fourier number: $Fo = \dfrac{k}{\rho c_p}\dfrac{t}{l^2}$

Prandtl number: $Pr = \dfrac{\mu c_p}{k}$

Froude number: $Fr = \dfrac{u}{\sqrt{lg}}$

Rayleigh number: $Ra = Gr\,Pr$

Grashof number: $Gr = \dfrac{g\beta\Delta T l^3 \rho^2}{\mu^2}$

Reynolds number (general): $Re = \dfrac{\rho u d}{\mu}$

Knudsen number: $Kn = \dfrac{\lambda}{l}$

Reynolds number (rotating disc): $Re = \dfrac{\rho\omega D^2}{4\mu}$

Mach number: $Ma = \dfrac{u}{a}$

Strouhal number: $St = \dfrac{nd}{u}$

Nusselt number: $Nu = \dfrac{hl}{k}$

Weber number: $We = u\sqrt{\dfrac{\rho l}{\sigma}}$

See also the specialist groups for hydraulic machines in section 9.6

9.10 *Pipe Friction Chart*

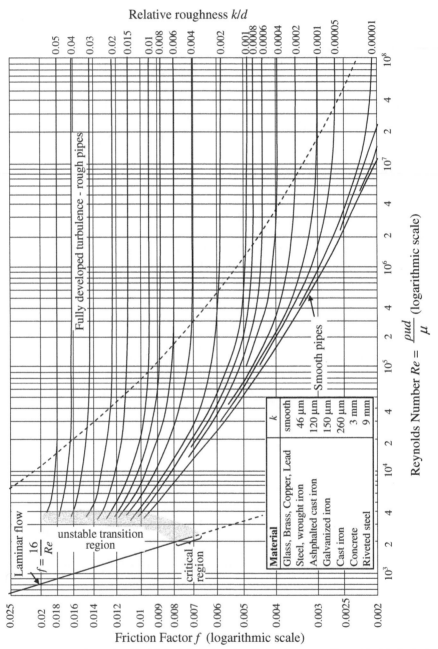

Note: the friction factor *f* used in this book is the Fanning friction factor, usually used by Mechanical and Chemical Engineers. Civil Engineers often use the alternative Darcy, Darcy-Weisbach, Moody or Blasius friction factor, which is four times larger.

9-10

10. Automatic Control
10.1 *Block Diagrams*

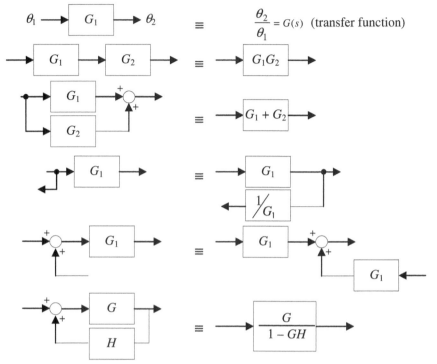

$\theta_1 \longrightarrow \boxed{G_1} \longrightarrow \theta_2 \quad\equiv\quad \dfrac{\theta_2}{\theta_1} = G(s) \text{ (transfer function)}$

$\boxed{G_1} \rightarrow \boxed{G_2} \quad\equiv\quad \boxed{G_1 G_2}$

$\equiv\quad \boxed{G_1 + G_2}$

$\equiv\quad \boxed{G_1} \;\; \boxed{1/G_1}$

$\equiv\quad \boxed{G_1} \;\; \boxed{G_1}$

$\equiv\quad \boxed{\dfrac{G}{1 - GH}}$

10.2 *Feedback Notation*

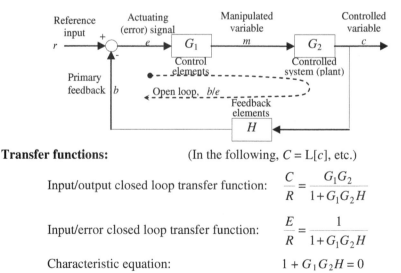

Reference input r — Actuating (error) signal e — G_1 Control elements — Manipulated variable m — G_2 Controlled system (plant) — Controlled variable c

Primary feedback b

Open loop, b/e

Feedback elements H

Transfer functions: (In the following, $C = L[c]$, etc.)

Input/output closed loop transfer function: $\dfrac{C}{R} = \dfrac{G_1 G_2}{1 + G_1 G_2 H}$

Input/error closed loop transfer function: $\dfrac{E}{R} = \dfrac{1}{1 + G_1 G_2 H}$

Characteristic equation: $1 + G_1 G_2 H = 0$

10-1

Open loop transfer function:

$$G_1 G_2 H \triangleq KGH = \frac{K \prod\limits_{i=1}^{m}(s+z_i)}{s^\ell \prod\limits_{i=1}^{n-\ell}(s+p_i)} \quad \begin{array}{ll} K & \text{gain} \\ z_i, p_i & \text{zeros and poles} \\ \ell & \text{system type number} \\ n & \text{system order number} \end{array}$$

10.3 *Stability Criteria for Linear Systems*

Root Location: No closed loop system pole may have a positive real part.

Routh Array:

Characteristic equation: $\quad a_n s^n + a_{n-1} s^{n-1} + \dots + a_1 s + a_0 = 0$

	1	2	3	...
n	a_n	a_{n-2}	a_{n-4}	...
$n-1$	a_{n-1}	a_{n-3}	a_{n-5}	...
$n-2$	b_1	b_2	b_3	...
$n-3$	c_1	c_2	c_3	...
...	...	...	...	...

$$b_1 = \frac{a_{n-1}a_{n-2} - a_n a_{n-3}}{a_{n-1}} \qquad\qquad b_2 = \frac{a_{n-1}a_{n-4} - a_n a_{n-5}}{a_{n-1}}$$

$$c_1 = \frac{b_1 a_{n-3} - a_{n-1} b_2}{b_1} \qquad\qquad c_2 = \frac{b_1 a_{n-5} - a_{n-1} b_3}{b_1}, \qquad \text{etc.}$$

Number of closed loop poles with positive real part = number of sign changes in column 1.

Nyquist encirclement:

$P = N + Z$

N = number of clockwise encirclements of $(-1, j0)$ by open loop locus

P = number of closed loop poles with positive real part

Z = number of open loop poles with positive real part

Gain Margin: $\quad |KG(j\omega_g)H(j\omega_g)|^{-1}$, for ω_g such that

$$\underline{/KG(j\omega_g)H(j\omega_g)} = -180°$$

Phase Margin: $\quad \underline{/KG(j\omega_p)H(j\omega_p)} + 180°$, for ω_p such that

$$|KG(j\omega_p)H(j\omega_p)| = 1$$

10.4 *Rules of Root Locus sketching*

1. Every point α on the root locus for positive K satisfies

$$\left| G(\alpha)H(\alpha) \right| = 1/K$$

$$\underline{/G(\alpha)K(\alpha)} = (1 + 2k) \times 180°, \qquad k = 0, \pm 1, \pm 2, \ldots$$

2. The number of branches of the root locus is equal to the number of poles.

3. Branches of the root locus can be considered to start on the poles ($K = 0$) and terminate on zeros ($K = \infty$) of the open loop system.

4. Points of the root locus exist on the real axis to the left of an odd number of poles plus zeros.

5. The locus is symmetrical with respect to the real axis.

6. The angles of asymptotes α_k to the root locus are given by

$$\alpha_k = \pm \frac{(2k+1)}{n-m} \pi, \qquad k = 0, 1, 2, \ldots$$

(n = number of poles, m = number of zeros)

7. The intersection of the asymptotes and the real axis occurs at

$$s_r = -\frac{\sum p_i - \sum z_i}{n-m}$$

8. The locus leaves the real axis or arrives on it at points α, given by

$$\frac{d}{d\alpha}[KG(\alpha)H(\alpha)] = 0$$

9. The intersection of the root locus and the imaginary axis can be found by application of Routh's stability criteria.

10.5 *Decibels*

Amplitude: $\qquad dB = 20 \log_{10}\left(\dfrac{V_1}{V_2}\right) \qquad \dfrac{V_1}{V_2} = 10^{dB/20}$

Power: $\qquad dB = 10 \log_{10}\left(\dfrac{W_1}{W_2}\right) \qquad \dfrac{W_1}{W_2} = 10^{dB/10}$

dB	Voltage ratio $\dfrac{V_1}{V_2}$	Power ratio $\dfrac{W_1}{W_2}$	dB	Voltage ratio $\dfrac{V_1}{V_2}$	Power ratio $\dfrac{W_1}{W_2}$
−40	0.01	0.0001	3	1.412	1.995
−30	0.0316	0.001	6	1.995	3.981
−20	0.1	0.01	10	3.162	10
−10	0.316	0.1	20	10	100
−6	0.501	0.251	30	31.62	1000
−3	0.708	0.501	40	100	10 000
0	1	1	50	316.2	100 000

10.6 *The Z-Transform*

Definition:
$$Z\{f(t)\} = F(z) = \sum_{k=0}^{\infty} f(kT)z^{-k}$$

The z-transform variable z is related to the Laplace transform variable s (see section 3.9) by $z = e^{sT}$

Properties:

$x(t)$	$X(z)$
$kx(t)$	$kX(z)$
$x_1(t) + x_2(t)$	$X_1(z) + X_2(z)$
$x(t + T)$	$zX(z) - zx(0)$
$tx(t)$	$-Tz\dfrac{dX(z)}{dz}$
$e^{-at}x(t)$	$X(ze^{at})$
$x(0)$ (initial value)	$\lim_{z \to \infty} X(z)$ (if the limit exists)
$x(\infty)$ (final value)	$\lim_{z \to 1} (z-1)X(z)$ (if the limit exists and the system is stable)

Transform Pairs

$x(t)$	$X(z)$
$\delta(t)$ $\begin{aligned} &= 1 \text{ for } t = 0 \\ &= 0 \text{ for } t = Kt \end{aligned}$	1
$\delta(t - kT)$ $\begin{aligned} &= 1 \text{ for } t = kT \\ &= 0 \text{ for } t \neq Kt \end{aligned}$	z^{-k}
$u(t)$ (unit step)	$\dfrac{z}{z-1}$
t	$\dfrac{Tz}{(z-1)^2}$
e^{-at}	$\dfrac{z}{(z-e^{-aT})}$
$1 - e^{-at}$	$\dfrac{(1-e^{-aT})z}{(z-1)(z-e^{-aT})}$
$\sin \omega t$	$\dfrac{z \sin \omega T}{(z^2 - 2z\cos \omega T + 1)}$
$\cos \omega t$	$\dfrac{z(z-\cos \omega T)}{(z^2 - 2z\cos \omega T + 1)}$
$e^{-at}\sin \omega t$	$\dfrac{(ze^{-at}\sin \omega T)}{(z^2 - 2ze^{-aT}\cos \omega T + e^{-2aT})}$
$e^{-at}\cos \omega t$	$\dfrac{(z^2 - ze^{-at}\cos \omega T)}{(z^2 - 2ze^{-aT}\cos \omega T + e^{-2aT})}$

10.7 *Table of Characteristic Systems*

Pole-Zero map and Root Locus	Frequency Response — Logarithmic (Bode)	Frequency Response — Complex Plane (Nyquist)	Step Response	Basic form of transfer function $G(s)$
Not applicable				K
				$\dfrac{1}{s}$
				$\dfrac{1}{1+Ts}$
				$\dfrac{1}{s(1+Ts)}$
$\phi = \cos^{-1}\xi$, $\omega_d = \omega_n\sqrt{1-\xi^2}$	$1/(2\xi\sqrt{(1-\xi^2)})$	$\omega = \omega_n$		$\dfrac{\omega_n^2}{s^2 + 2\xi\omega_n s + \omega_n^2}$
				$\dfrac{1+\alpha Ts}{1+Ts}$
	$\omega_n/(2\xi)$			$\dfrac{\omega_n^2(1+2\xi s/\omega_n)}{s^2 + 2\xi\omega_n s + \omega_n^2}$

Typical Example: Dynamic, Electrical

10-5

11. Electricity and Magnetism
11.1 *Basic Relationships*

Ohm's Law: $V = IR$ $\qquad$ $R = \dfrac{V}{I}$ $\qquad$ $I = \dfrac{V}{R}$

Power: $\qquad$ DC: $P = VI = I^2R = \dfrac{V^2}{R}$ $\quad$ AC: $P = \mathrm{Re}(\underline{V} \bullet \underline{I}^*) = |V||I|\cos\phi$

Resistance: $I = \dfrac{a}{\rho_0(1+\alpha(T-T_0))}\dfrac{\mathrm{d}V}{\mathrm{d}x}$ $\qquad$ $R = \displaystyle\int \dfrac{\rho_0(1+\alpha(T-T_0))}{a}\mathrm{d}x$

Inductance:

$$E = -L\dfrac{\mathrm{d}I}{\mathrm{d}t} \qquad I = -\int\dfrac{V}{L}\mathrm{d}t \qquad L = N^2\mu_0\mu_r\dfrac{a}{\ell}$$

$\qquad$ Stored energy $= \tfrac{1}{2}LI^2$ $\quad$ L R circuit decay: $\qquad$ $I = I_0 e^{-Rt/L}$

Capacitance:

$$Q = CV = \int I\mathrm{d}t \qquad\qquad I = \dfrac{\mathrm{d}Q}{\mathrm{d}t} = C\dfrac{\mathrm{d}V}{\mathrm{d}t}$$

$$C = \varepsilon_0\varepsilon_r(n-1)\dfrac{a}{d} \text{ for } n \text{ parallel plates}$$

$\qquad$ Stored energy $= \tfrac{1}{2}CV^2$ $\qquad$ $F = \tfrac{1}{2}\varepsilon_0\varepsilon_r a\left(\dfrac{V}{x}\right)^2$

Electrostatics:

$$F = \dfrac{Q_1 Q_2}{4\pi\varepsilon_0 r^2} \qquad\qquad \underline{F} = e\underline{E} = -e\,\mathrm{grad}\,V$$

$$Q = \oint \underline{D} \bullet \mathrm{d}\underline{S} \qquad (= \Psi) \quad \underline{D} = \varepsilon_0\varepsilon_r\underline{E}$$

Electromagnetism:

Maxwell's Equations:

$$\nabla \bullet \mathbf{D} = \rho \quad \nabla \bullet \mathbf{B} = 0 \quad \nabla \times \mathbf{H} - \dfrac{\partial \mathbf{D}}{\partial t} = \mathbf{J} \quad \nabla \times \mathbf{E} + \dfrac{\partial \mathbf{B}}{\partial t} = 0$$

$$E = -N\dfrac{\mathrm{d}\Phi}{\mathrm{d}t} \qquad B = \dfrac{\mu_0 I}{2\pi r}$$

$F = B\ell I$
(perpendicular
components only)
$\qquad F = \dfrac{\mu_0 I_1 I_2 \ell}{2\pi d}$

$\dfrac{\mathrm{d}H}{\mathrm{d}\ell} = \dfrac{I\sin\alpha}{4\pi x^2}$ $\quad$ for solenoid: $H = \dfrac{NI}{\ell}$

Fleming's RH rule

Magnetism

$$H = \frac{B}{\mu_0 \mu_r}. \text{ For a magnetic circuit, } B = \frac{\Phi}{a}; \qquad \Phi = \frac{NI}{\dfrac{\ell_1}{\mu_1 a_1} + \dfrac{\ell_2}{\mu_2 a_2}}$$

$$\text{Stored energy density} = \frac{1}{2} HB = \frac{1}{2} \frac{B^2}{\mu_0}, \qquad F = \left(\frac{1}{2} HB\right) a = \frac{B^2 a}{2\mu_0}$$

AC Circuits

$$V_{rms} = \frac{1}{\sqrt{2}} V_{max} \qquad \qquad \omega = 2\pi f$$

Series LCR:

$$Z = \left[R^2 + \left(\omega L - \frac{1}{\omega C}\right)^2 \right]^{0.5}$$

$$\underline{Z} = R + j\omega L + \frac{1}{j\omega C} = R + j\left(\omega L - \frac{1}{\omega C}\right) \qquad \cos\phi = \frac{R}{Z}$$

$$\text{At resonance,} \qquad \omega = \omega_o = \frac{1}{\sqrt{LC}} \qquad \qquad Q \text{ factor} = \omega_o \frac{L}{R}$$

Star-Delta Transformation

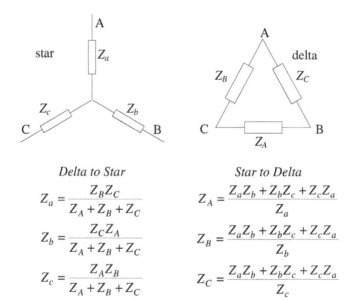

Delta to Star

$$Z_a = \frac{Z_B Z_C}{Z_A + Z_B + Z_C}$$

$$Z_b = \frac{Z_C Z_A}{Z_A + Z_B + Z_C}$$

$$Z_c = \frac{Z_A Z_B}{Z_A + Z_B + Z_C}$$

Star to Delta

$$Z_A = \frac{Z_a Z_b + Z_b Z_c + Z_c Z_a}{Z_a}$$

$$Z_B = \frac{Z_a Z_b + Z_b Z_c + Z_c Z_a}{Z_b}$$

$$Z_C = \frac{Z_a Z_b + Z_b Z_c + Z_c Z_a}{Z_c}$$

11.2 *Machines*

AC Machines

Synchronous speed $= f/p$

$E \propto f\Phi$ (rms)

$$T \propto \frac{\Phi^2 sR}{R^2 + (s X_0)^2}$$

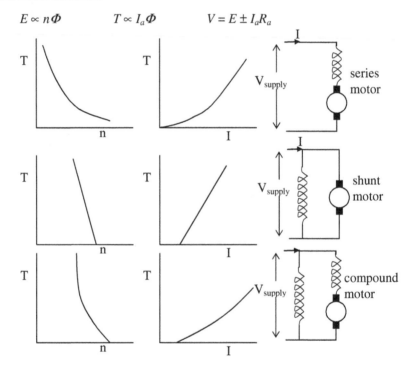

DC/Universal machine:

$E \propto n\Phi$ $\qquad$ $T \propto I_a\Phi$ $\qquad$ $V = E \pm I_a R_a$

Components:

Colour Code

0	Black	2	Red	4	Yellow	6	Blue	8	Grey
1	Brown	3	Orange	5	Green	7	Violet	9	White

Preferred Values

10, 12, 15, 18, 22, 27, 33, 39, 47, 56, 68, 82

11.3 *Operational Amplifiers*

Basic Circuits

$$V_o \approx -\frac{R_2}{R_1} V_i$$

Inverting Amplifier

$$V_o \approx (1+\frac{R_2}{R_1})V_i$$

Non -inverting amplifier

$$V_o \approx \frac{R_2}{R_1}(V_2 - V_1)$$
with $\frac{R_1}{R_2} = \frac{R_3}{R_4}$

Differential Input Amplifier

$$V_o = -(k_1 V_1 + k_2 V_2 + ...)$$

Adding (Inverting)/Scaling

$$V_o \approx V_i$$

Voltage follower

$$V_o = IR$$

Current to voltage converter

$$V_o \approx -\frac{1}{RC}\int_0^t V_i \, dt$$

Integrating amplifier

$$V_o \approx -RC\frac{dV_i}{dt}$$

Differentiating amplifier

μA741 Op Amp Characteristics

11-4

11.4 *The Electromagnetic Spectrum*

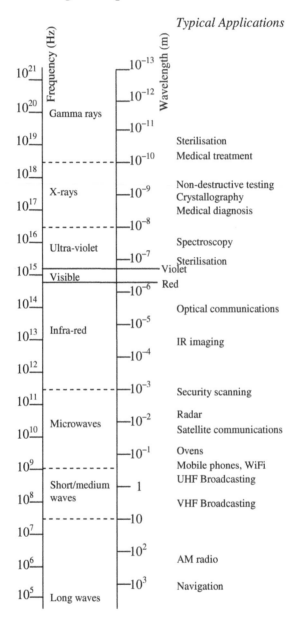

12. Soil Mechanics
12.1 *Soil Classification*

Particle size (see also Grain Size of Clastic Sedimentary Rocks, section 8.4)

										Size mm
0.002	0.006	0.02	0.06	0.2	0.6	2	6	20	60	200

F = Fine, M = Medium, C = Coarse

Casagrande Soil Classification for fine grained soils

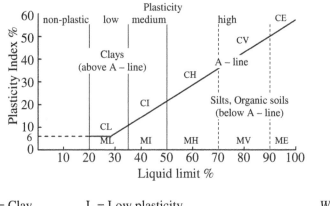

C = Clay   L = Low plasticity $W_{LL} < 35\%$

M = Silt   I = Intermediate (medium) plasticity $35\% \le W_{LL} < 50\%$

O = Organic   H = High plasticity $50\% \le W_{LL} < 70\%$

Pt = Peat   V = Very high plasticity $70\% \le W_{LL} < 90\%$

     E = Extremely high plasticity $W_{LL} \ge 90\%$

12.2 *Volume – weight relationships*

Definitions:

Void ratio $\qquad e = \dfrac{v_v}{v_s}$ $\qquad\qquad$ Specific volume $\quad v = \dfrac{v_t}{v_s} = 1 + e$

Porosity $\qquad n = \dfrac{v_v}{v_t} = \dfrac{e}{1+e}$ $\qquad$ Saturation ratio $\quad S_r = \dfrac{v_w}{v_v}$

Water content $\quad w = \dfrac{m_w}{m_s}$ $\qquad\qquad$ Relative density $\quad D_r = \dfrac{e_{max} - e}{e_{max} - e_{min}}$

$wG_s = eS_r$ $\qquad\quad \gamma = \dfrac{G_s \gamma_w (1+w)}{v}$ $\qquad\qquad \gamma_{dry} = \dfrac{G_s \gamma_w}{v}$

12.3 *Stratigraphic Table*

A table of the geological timescale is given in section 8.2

12.4 *Effective Stress and Seepage*

$$\sigma \quad = \quad \sigma' \quad + \quad u$$

Total stress $\;=\;$ Effective stress $\;+\;$ pore water pressure

Hydraulic gradient: $\qquad\qquad\qquad i = -\dfrac{dh}{dx}$

Critical hydraulic gradient: $\qquad\quad i_{crit} = \dfrac{\gamma - \gamma_w}{\gamma_w}$

Darcy's Law: $\qquad\qquad\qquad\qquad q = Aki \text{ or } v = ki$

Flownet analysis: $\qquad\qquad\qquad q = kH\,\dfrac{N_f}{N_h} \qquad$ per metre length or thickness

Transformation factor in anisotropic soils: $\quad x' = \sqrt{\dfrac{k_y}{k_x}}$

Transformed permeability: $\qquad\quad k_t = \sqrt{k_x k_y}$

Deflection of flowlines at interface: $\dfrac{\tan \beta_1}{\tan \beta_2} = \dfrac{k_1}{k_2}$

12.5 *Stress Parameters*

Average effective stress: $p' = \frac{1}{3}(\sigma_1' + \sigma_2' + \sigma_3')$

Deviator stress:

$$q = \frac{1}{\sqrt{2}}\left\{(\sigma_1' - \sigma_2')^2 + (\sigma_1' - \sigma_3')^2 + (\sigma_2' - \sigma_3')^2\right\}^{\frac{1}{2}}$$

$$= (\sigma_1' - \sigma_3') \quad \text{if} \quad \sigma_1' = \sigma_2' \quad \text{or} \quad \sigma_2' = \sigma_3' \qquad \text{(triaxial conditions)}$$

Stress ratio: $\eta = \dfrac{q}{p'}$

Plane strain: $s' = \frac{1}{2}(\sigma_1 + \sigma_3)$ $t = \frac{1}{2}(\sigma_1 - \sigma_3)$

12.6 *Failure Criteria*

In terms of effective stresses: $\left(\dfrac{\tau}{\sigma'}\right)_{max} = \tan\phi'$

In terms of total stresses: $\tau_{max} = \tau_u$ (undrained shear strength)

12.7 *Consolidation-time curves*

Curve (1)   (1)   (2)   (3)   (4)

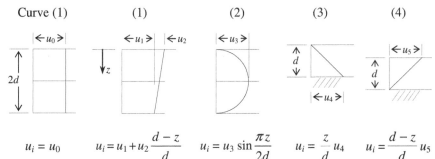

$$u_i = u_0 \qquad u_i = u_1 + u_2\frac{d - z}{d} \qquad u_i = u_3\sin\frac{\pi z}{2d} \qquad u_i = \frac{z}{d}u_4 \qquad u_i = \frac{d - z}{d}u_5$$

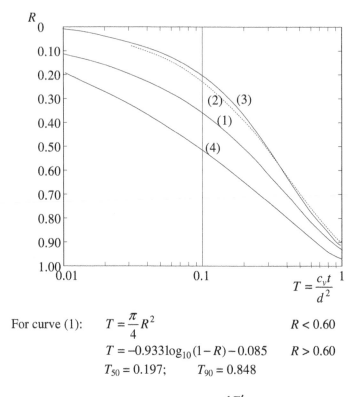

For curve (1): $\quad T = \dfrac{\pi}{4}R^2$ $\qquad\qquad\qquad\qquad$ $R < 0.60$

$\qquad\qquad\quad T = -0.933\log_{10}(1-R) - 0.085$ $\qquad R > 0.60$

$\qquad\qquad\quad T_{50} = 0.197; \qquad T_{90} = 0.848$

Consolidation coefficient: $\qquad\qquad c_v = \dfrac{kE_0'}{\gamma_w}$

t_x method for calculating c_v from oedometer test data:

$\qquad$ plot settlement ρ against $\sqrt{\text{time}}$, and $t_x = \dfrac{3d^2}{4c_v}$

12.8 *Cam clay*

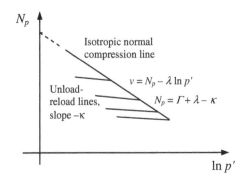

12.9 *Stresses and Displacements in Elastic Half-space*

Note: $m = L/Z$
$n = B/Z$

m and n are interchangeable
$\sigma_z = q\, I_\sigma$

Influence factors for the increase in vertical stress at a depth Z below the corner of a uniformly loaded rectangle

12-5

Boussinesq

Point load Q at surface $\longleftrightarrow$ *Line load q at surface*

vertical stress: $\sigma_z = \dfrac{3Q}{2\pi z^2}\cos^5\theta$ $\qquad$ $\sigma_z = \dfrac{2q}{\pi z}\cos^4\theta$

surface settlement: $w_s = \dfrac{Q(1+v)}{2\pi z E}\cos\theta[\cos^2\theta + 2(1-v)]$ $\quad w_s = \dfrac{2q(1-v^2)}{\pi E}\ln\left(\dfrac{d}{x}\right)$

where displacement at d is assumed $= 0$ $(d \ge x$

12.10 *Retaining walls and lateral earth pressure*

$\sigma_h' = K_o\sigma_v'$ $\qquad$ in situ

$\sigma_h' = K_a\sigma_v'$ $\qquad$ active

$\sigma_h' = K_p\sigma_v'$ $\qquad$ passive

For frictionless walls, $K_a = \dfrac{1-\sin\phi'}{1+\sin\phi'}$; $\quad K_p = \dfrac{1+\sin\phi'}{1-\sin\phi'}$

Total stress analysis: $\qquad \sigma_h = \sigma_v - 2\,\tau_u$ $\qquad$ (active)

$\qquad\qquad\qquad\qquad\quad \sigma_h = \sigma_v + 2\,\tau_u$ $\qquad$ (passive)

Tension crack depth: $\qquad\qquad\qquad\qquad\qquad$ (dry)

(surface surcharge q) $\qquad z = \dfrac{2\tau_u - q}{\gamma}$

$\qquad\qquad\qquad\qquad\qquad z = \dfrac{2\tau_u - q}{\gamma - \gamma_w}$ $\qquad$ (flooded)

12.11 *Bearing capacity for vertical loads*

Effective stress analysis

Bearing capacity σ_f: foundation length L, breadth B, depth D, failure criterion $(\tau/\sigma)_{max} = \tan\phi'$:

$$\sigma_f' = \{N_q s_q d_q \sigma_o'\} + \left\{N_\gamma s_\gamma d_\gamma r_\gamma\left(\dfrac{\gamma B}{2} - \Delta u\right)\right\}$$

where $N_q = K_p\, e^{\pi\tan\phi'}$ $\qquad r_\gamma = 1 - 0.25\log_{10}(B/2)$ for $B \ge 2$ m

Parameter	Meyerhof $(\phi' > 10°)$	Brinch Hansen
Shape factor s_q	$1 + 0.1\ K_p\ (B/L)$	$1 + [(B/L)\tan\phi']$
Depth factor d_q	$1 + 0.1\ \sqrt{K_p}\ (D/B)$	$1 + 2\tan\phi'\ (1 - \sin\phi')k$
		$k = D/B \qquad\qquad D/B \leq 1$
		$k = \tan^{-1}(D/B) \qquad D/B > 1$
		(radians)
N_γ	$(N_q - 1)\tan(1.4\phi')$	$1.5\ (N_q - 1)\tan\phi'$
Shape factor s_γ (for N_γ) $= s_q$		$1 - 0.4\ (B/L)$
Depth factor d_γ (for N_γ) $= d_q$		1

Total stress analysis

Bearing capacity σ_f: foundation length L, breadth B, depth D, failure criterion $\tau_{\max} = \tau_u$:

$$(\sigma_f - \sigma_o) = N_c s_c d_c \tau_u \quad \text{where} \quad N_c = 5.14$$

Parameter	Skempton	Meyerhof
Shape factor s_c	$1 + 0.2\ (B/L)$	$1 + 0.2\ (B/L)$
Depth factor d_c	$1 + 0.23\ \sqrt{(D/B)}$ up to a maximum of $1.46\ (D/B = 4)$	$1 + 0.2\ (D/B)$

12.12 *Slope Stability*

Pore pressure ratio: $r_u = \dfrac{u}{\gamma h}$

$$\tau_{mob} = \frac{\tau_u}{F} \qquad \text{(total stress analysis)}$$

$$\tan\phi'_{mob} = \frac{\tan\phi'}{F} \qquad \begin{array}{l}\text{(effective stress analysis)}\\ \text{(F = factor of safety)}\end{array}$$

Circular slips:

$$F = \frac{R\sum(\tau_u l)}{\sum x_w w} \qquad \text{(total stress analysis)}$$

$$F = \frac{\sum[(w\cos\alpha - ul)\tan\phi']}{\sum w\sin\alpha} \qquad \text{(effective stress analysis)}$$

Bishop's simplified method:

$$F = \frac{1}{\sum w\sin\alpha}\sum\left[\frac{(w - ub)\tan\phi'}{\cos\alpha + \dfrac{\tan\phi'\sin\alpha}{F}}\right]$$

13. Structures

The section second moments of area required in many formulae in this section may be found from the formulae in Section 5.1.

13.1 *Elastic bending and torsion relationships*

For elastic, isotropic materials,

$$G = \frac{E}{2(1+v)} \qquad K = \frac{E}{3(1-2v)}$$

Hooke's law in two dimensions:

$$\varepsilon_x = \frac{1}{E}(\sigma_x - v\sigma_y) \qquad \varepsilon_y = \frac{1}{E}(\sigma_y - v\sigma_x)$$

Simple bending: $\dfrac{M}{I} = \dfrac{\sigma}{y} = \dfrac{E}{R}$ $\qquad \sigma = \dfrac{M}{Z} \qquad Z = \dfrac{I}{h_{max}}$

$$M = -EI\frac{d^2y}{dx^2} \qquad Q = \frac{dM}{dx}$$

Shear stress in beam: $\qquad \tau = \dfrac{QA\overline{y}}{bI}$

Torsion of a circular section: $\qquad \dfrac{T}{J} = \dfrac{\tau}{r} = \dfrac{G\theta}{L}$

For a solid section, $\qquad J = \dfrac{\pi d^4}{32} = A\dfrac{d^2}{8}$

For a tube, $\qquad J = \dfrac{\pi}{32}(d_2^4 - d_1^4) = \dfrac{A}{8}(d_2^2 + d_1^2)$

(d_1 = inner diameter, d_2 = outer diameter)

13.2 *Buckling Loads*

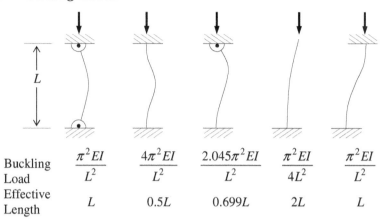

Buckling Load	$\dfrac{\pi^2 EI}{L^2}$	$\dfrac{4\pi^2 EI}{L^2}$	$\dfrac{2.045\pi^2 EI}{L^2}$	$\dfrac{\pi^2 EI}{4L^2}$	$\dfrac{\pi^2 EI}{L^2}$
Effective Length	L	$0.5L$	$0.699L$	$2L$	L

13.3 *Beams bent about principal axis*

L = overall length W = point load, M = moment w = load per unit length	End Slope	Max Deflection	Max bending moment
(cantilever with moment M)	$\dfrac{ML}{EI}$	$\dfrac{ML^2}{2EI}$	M
(cantilever with point load W)	$\dfrac{WL^2}{2EI}$	$\dfrac{WL^3}{3EI}$	WL
(cantilever with distributed load w)	$\dfrac{wL^3}{6EI}$	$\dfrac{wL^4}{8EI}$	$\dfrac{wL^2}{2}$
(simply supported with moments M, M)	$\dfrac{ML}{2EI}$	$\dfrac{ML^2}{8EI}$	M
(simply supported, point load W at ½ L, ½ L)	$\dfrac{WL^2}{16EI}$	$\dfrac{WL^3}{48EI}$	$\dfrac{WL}{4}$
(simply supported, distributed load w)	$\dfrac{wL^3}{24EI}$	$\dfrac{5wL^4}{384EI}$	$\dfrac{wL^2}{8}$
A $\downarrow W$ $\leftarrow c \rightarrow B$ $\leftarrow a \rightarrow \mid \leftarrow b \longrightarrow$ $a \le b,\ \ c = \sqrt{\tfrac{1}{3} b(L+a)}$	$\theta_B = \dfrac{Wac^2}{2LEI}$ $\theta_A = \dfrac{L+b}{L+a}\,\theta_B$	$\dfrac{Wac^3}{3LEI}$ (at position c)	$\dfrac{Wab}{L}$ (under load)

13.4 *Fixed End Shear and Moments*

L = overall length, W = point load, M = moment, w = load per unit length				Maximum deflection Δ	Max deflection position c (from RH end)	Max bending moment (modulus)
LH End		RH End				
moment	shear	shear	moment			
$\dfrac{wL^2}{12}$	$\tfrac{1}{2}wL$; w	$\tfrac{1}{2}wL$	$\dfrac{wL^2}{12}$	$\dfrac{wL^4}{384EI}$	$\dfrac{L}{2}$	$\dfrac{wL^2}{12}$
$\dfrac{WL}{8}$	$\tfrac{1}{2}W$; W ; $\leftarrow 0.5L \rightarrow$	$\tfrac{1}{2}W$	$\dfrac{WL}{8}$	$\dfrac{WL^3}{192EI}$	$\dfrac{L}{2}$	$\dfrac{WL}{8}$
$\dfrac{Wab^2}{L^2}$	$\dfrac{Wb^2(L+2a)}{L^3}$; W ; $\leftarrow a \rightarrow \leftarrow b \longrightarrow$; $\leftarrow c \rightarrow$	$\dfrac{Wa^2(L+2b)}{L^3}$	$\dfrac{Wa^2b}{L^2}$	$\dfrac{Wa^2bc^2}{6EIL^2}$	$\dfrac{2Lb}{L+2b}$ a ≤ b	$\dfrac{Wab^2}{L^2}$
$\dfrac{6EI\delta}{L^2}$	$\dfrac{12EI\delta}{L^3}$	$\dfrac{12EI\delta}{L^3}$; δ	$\dfrac{6EI\delta}{L^2}$	δ	0	$\dfrac{6EI\delta}{L^2}$
$\dfrac{Mb}{L^2}(2a-b)$	$\dfrac{M}{L}$; M ; $\leftarrow a \rightarrow \leftarrow b \longrightarrow$	$\dfrac{M}{L}$	$\dfrac{Ma}{L^2}(2b-a)$	-	-	$\dfrac{M}{L^2}a(2b-a)$
$\dfrac{wL^2}{30}$	$\dfrac{3wL}{20}$; w	$\dfrac{7wL}{20}$; $\leftarrow c \rightarrow$	$\dfrac{wL^2}{20}$	$\dfrac{wL^4}{764EI}$	$0.475\,L$	$\dfrac{wL^2}{20}$
$\dfrac{3WL}{16}$	$\dfrac{11W}{16}$; W ; $\leftarrow c \rightarrow$; $\leftarrow 0.5L \rightarrow$	$\dfrac{5W}{16}$; prop 0		$\dfrac{2WL^3}{215EI}$	$0.447\,L$	$\dfrac{3WL}{16}$
$\dfrac{Wab(L+b)}{2L^2}$ $= M$	$\dfrac{Wb+M}{L}$; W ; $\leftarrow a \rightarrow \leftarrow b \longrightarrow$; $\leftarrow c \rightarrow$	$\dfrac{Wa-M}{L}$; 0 prop		$\dfrac{Wa^2bc}{6EIL}$	$L\sqrt{\dfrac{b}{2L+b}}$ b ≥ 0.4142 L	$\dfrac{Wab(L+b)}{2L^2}$
$\dfrac{wL^2}{8}$	$\dfrac{5wL}{8}$; w	$\dfrac{3wL}{8}$; $\leftarrow c \rightarrow$	0 prop	$\dfrac{wL^4}{185EI}$	$0.442L$	$\dfrac{wL^2}{8}$

13.5 *Beam stiffness and coefficients*

In the following, F_i represent axial or shear forces or bending or torsional couples, according to context; x_i represent linear or rotational displacements as appropriate.

All beam and frame stiffness matrices may be built up from the following orthogonal components for each beam element:

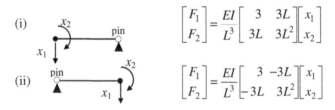

(a) axial:

$$\begin{bmatrix} F_1 \\ F_2 \end{bmatrix} = \frac{EA}{L} \begin{bmatrix} 1 & -1 \\ -1 & 1 \end{bmatrix} \begin{bmatrix} x_1 \\ x_2 \end{bmatrix}$$

(b) torsional:

$$\begin{bmatrix} F_1 \\ F_2 \end{bmatrix} = \frac{GJ}{L} \begin{bmatrix} 1 & -1 \\ -1 & 1 \end{bmatrix} \begin{bmatrix} x_1 \\ x_2 \end{bmatrix}$$

(c) plane bending:

$$\begin{bmatrix} F_1 \\ F_2 \\ F_3 \\ F_4 \end{bmatrix} = \frac{EI}{L^3} \begin{bmatrix} 12 & -12 & 6L & 6L \\ -12 & 12 & -6L & -6L \\ 6L & -6L & 4L^2 & 2L^2 \\ 6L & -6L & 2L^2 & 4L^2 \end{bmatrix} \begin{bmatrix} x_1 \\ x_2 \\ x_3 \\ x_4 \end{bmatrix}$$

Other forms of (c) arise if one end is pinned and vertically constrained:

(i)

$$\begin{bmatrix} F_1 \\ F_2 \end{bmatrix} = \frac{EI}{L^3} \begin{bmatrix} 3 & 3L \\ 3L & 3L^2 \end{bmatrix} \begin{bmatrix} x_1 \\ x_2 \end{bmatrix}$$

(ii)

$$\begin{bmatrix} F_1 \\ F_2 \end{bmatrix} = \frac{EI}{L^3} \begin{bmatrix} 3 & -3L \\ -3L & 3L^2 \end{bmatrix} \begin{bmatrix} x_1 \\ x_2 \end{bmatrix}$$

In all cases, if one end is fixed and considered as a reaction, its forces and deflections may be ignored with a corresponding reduction in the stiffness matrix.

The three components (a), (b) and (c) may be combined into larger matrices with zeros in all unspecified positions.

A general plane frame element will have components (a) and (c), and require a 6×6 stiffness matrix. A general space frame element will have components (a) and (b), and two components of (c), for two different planes of bending. It will require a 12×12 stiffness matrix. In general, the two components of (c) will have different values of I.

13.6 *Shear*

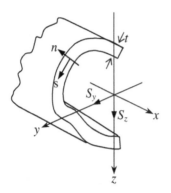

Shear flow per unit length of wall resulting from the *applied* shear forces S_z, S_y is

$$q = t\tau_{xs} = \frac{(-S_z)}{I_{yy}I_{zz} - I_{yz}^2}\left(I_{zz}\int_A z\,dA - I_{yz}\int_A y\,dA\right)$$

$$+ \frac{(-S_y)}{I_{yy}I_{zz} - I_{yz}^2}\left(I_{yy}\int_A y\,dA - I_{yz}\int_A z\,dA\right)$$

Resultant force for this shear flow acts through the *Shear Centre*.

13.7 *Torsion*

For a thin walled closed section,

$$T_x = 2Aq = \frac{4A^2G}{\displaystyle\int_0^s \frac{1}{t}\,ds} \times \frac{\theta}{L}$$

A = enclosed area

For a thin rectangular section,

$$T_x = \frac{dt^2}{3}\tau_{zx\,max} = \frac{dt^3G}{3}\frac{\theta}{L}$$

(For torsion of a circular rod or tube, see page 13.1)

13.8 *Asymmetric bending*

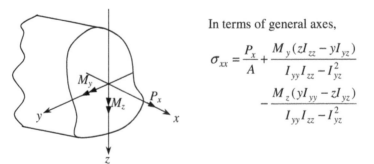

In terms of general axes,

$$\sigma_{xx} = \frac{P_x}{A} + \frac{M_y(zI_{zz} - yI_{yz})}{I_{yy}I_{zz} - I_{yz}^2}$$

$$- \frac{M_z(yI_{yy} - zI_{yz})}{I_{yy}I_{zz} - I_{yz}^2}$$

When the principal axes *m*, *n* lie in directions *y* and *z*, I_{yz} will be zero, and

$$\sigma_{xx} = \frac{P}{A} + \frac{nM_n}{I_{mm}} - \frac{mM_n}{I_{nn}}$$

13-5

13.9 *Stress and strain transformations*

Mohr's circle for two-dimensional stress system

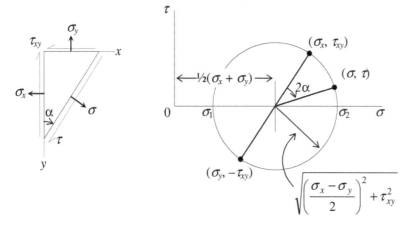

Mohr's circle for two-dimensional strain

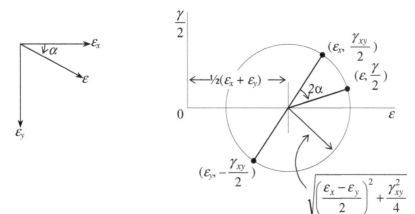

ε_x, ε_y, and ε correspond to σ_x, σ_y and σ, respectively.

$\dfrac{\gamma_{xy}}{2}$ and $\dfrac{\gamma}{2}$ correspond to τ_{xy} and τ, respectively.

Three-dimensional stress system

Principal direct stresses σ_1, σ_2 and σ_3 are equivalent to principal shear stresses $\tfrac{1}{2}(\sigma_1 - \sigma_2)$, $\tfrac{1}{2}(\sigma_2 - \sigma_3)$ and $\tfrac{1}{2}(\sigma_3 - \sigma_1)$.

Strain energy per unit volume:

$$U = \frac{(\sigma_1 + \sigma_2 + \sigma_3)}{18K} + \frac{\left[(\sigma_1 - \sigma_2)^2 + (\sigma_2 - \sigma_3)^2 + (\sigma_3 - \sigma_1)^2\right]}{12G}$$

14. Symbols Index

N.B. Symbols with very clear conventional meanings, symbols which are adequately defined where they appear in the text, symbols locally modified with subscripts and symbols which represent constants in a particular context are omitted from this list. Units given are the basic units, without the multipliers which may be used in practice.

Symbol	Section	Quantity	SI unit
a	5.2	acceleration	ms^{-2}
	9.8	velocity of sound	ms^{-1}
	11.1	area	m^2
A	5.1	area	m^2
	6.1	atomic mass	-
B	11.1	magnetic flux density	T
c	9.2	velocity	ms^{-1}
c_p, c_v	ch 7, 9	specific heat at constant pressure, volume	$Jkg^{-1}K^{-1}$
c_v	12.7	consolidation coefficient	m^2s^{-1}
C	6.5	concentration	$mol\ m^{-2}$
	9.7	Chézy coefficient	$m^{1/2}s^{-1}$
	11.1, 11.3	capacitance	F
d	6.1	interatomic spacing	m
	9.7	depth	m
D	6.5	diffusion coefficient	m^2s^{-1}
	8.6	dynamic or geopotential height	Jkg^{-1}
e	12.2	void ratio	-
E	ch 7, ch 13	Young's modulus	Nm^{-2}
	11.1, 11.2	electromotive force (emf)	V
ΔE	6.1	energy difference	J
f	9.4	(Fanning) friction factor	-
	11.1	frequency	Hz
F	many	force, tension, generalised force	N
g	many	gravitational intensity (acceleration)	$Nkg^{-1}\ (ms^{-2})$
G	ch 7, 13.1, 13.9	shear (rigidity) modulus	Nm^{-2}
	12.2		
G_s		relative density of solids	-
h	9.3, 9.9	heat transfer coefficient	$Wm^{-2}K^{-1}$
	9.4	bearing clearance	m
h, h_{fg}	9.2	specific enthalpy, of evaporation	Jkg^{-1}
H	ch 9	enthalpy	J
	9.6	head	m
	11.1	magnetic field strength	Am^{-1}
i	12.4	hydraulic gradient	-
I, I_{nn}	ch 5, ch 13	second moment of area (about axis nn)	m^4
		moment of inertia (about axis nn)	$kg\ m^2$
I	ch 11	current	A
J	5.3, 13.1	polar moment of inertia	$kg\ m^2$
	6.5	diffusion flux	$kgm^{-2}s^{-1}$

14.1

Symbol	Section	Quantity	SI unit
k, k_{nn}	ch 5, 9.4	radius of gyration (about axis nn)	m
k	6.5	Boltzmann's constant	JK^{-1}
	ch 7, 9.9	thermal conductivity	$Wm^{-1}K^{-1}$
	9.6	cavitation number	-
K	ch 7, 13.1, 13.9	bulk modulus	Nm^{-2}
K, K_{IC},	6.6, ch 7	stress intensity factor	$Nm^{-3/2}$
ΔK	6.6	stress intensity range	
K_p	9.2	equilibrium constant	$Pa^{1/n}$
l	many	length	m
ℓ	11.1	total length of conductor	m
L	many	length	m
	11.1	inductance	H
m	many	mass	kg
	9.7	hydraulic mean depth (radius)	m
M	many	mass	kg
	9.2	molecular mass	-
	6.6, ch 13	(bending) moment	Nm
n	6.1	atoms per unit cell	-
	9.2, 9.8	polytropic index	-
	9.7	Manning roughness coefficient	-
	9.9, 11.2	frequency, rotational speed	Hz, s^{-1}
	12.2	porosity	-
N	5.2	normal force	N
	6.1	Avogadro's number	-
	6.4	quadratic Miller indices	-
N'	11.1	number of turns	-
N_p	6.5	total number of atoms	-
N_p	9.2	number of power cycles/unit time	s^{-1}
N_i	6.6	total cycles to failure at strain amplitude	-
	6.6	total cycles to failure at stress amplitude	-
p, Δp	ch 9	pressure (difference)	Pa
p_v, p_∞	9.6	vapour, reference (free stream) pressure	Pa
p_0	9.8	stagnation pressure	Pa
p	11.2	no of pole pairs	-
p'	12.5	average effective stress	Nm^{-2}
P	9.4	total (stagnation) pressure	Pa
P_x	13.8	axial force	N
q	12.4	flow rate	$m^3 s^{-1}$
	12.5	deviator stress	Nm^{-2}
q_b	9.3	emissive power of a black body	Wm^{-2}
Q	ch 9	heat	J
	ch 9	volume flow rate, discharge	$m^3 s^{-1}$
	6.5	activation energy	J
	6.6	(shear) force, load	N
	11.1	charge	C

Symbol	Section	Quantity	SI unit
R	7.3	gas constant	$Jkg^{-1}K^{-1}$
	9.7	hydraulic mean radius (depth)	m
	ch 11	resistance	Ω
	13.1	radius of curvature	m
R_0	6.5	universal gas constant	$Jmol^{-1}K^{-1}$
s	11.2	fractional slip	-
s, s_{fg}	ch 9,(9.2)	specific entropy (of evaporation)	$Jkg^{-1}K^{-1}$
S	ch 9	entropy	JK^{-1}
	8.6	salinity	-
S, S_f, S_0	9.7	channel, friction, invert slope	-
S_y, S_z	13.6	shear force	N
$t_{1/2}$	8.4	Radioactive half life	s
T	5.2, 9.6	torque	Nm
	13.1, 13.7		
T, ΔT	many	temperature, temperature difference	K
T_0	many	stagnation/reference temperature	K
T_c	7.3	critical temperature	K
u	5.2	initial velocity	ms^{-1}
	12.4	pore water pressure	Pa
u, u_m, u_∞	ch 9	velocity, free stream (external) velocity	ms^{-1}
u_τ		"friction velocity"	ms^{-1}
u, u_{fg}	9.2	specific internal energy, of evaporation	Jkg^{-1}
U	ch 9	internal energy	J
v	5.2, 12.4	velocity	ms^{-1}
	9.2, 12.2	specific volume	m^3kg^{-1}
	12.2	volume	m^3
v_∞	9.6	reference (free stream) velocity	ms^{-1}
v_s	9.2	swept volume	m^3
V	6.1	volume of unit cell	m^3
	ch 9	volume	m^3
	ch 11	voltage	V
w	12.2	water content	-
	ch 13	load/unit length	Nm^{-1}
W	ch 9	work	J
	10.5	power	W
	ch 13	load	N
X_0	11.2	leakage reactance per phase	Ω
z	9.2, 9.4	height, head, elevation	m
	10.6	z-transform variable	
Z	9.4	total head	m
Z, $\underline{Z}$	11.1	impedance, complex	Ω
α	6.8, ch 7	coefficient of thermal linear expansion	K^{-1}
	8.4	isotope fractionation factor	-
	8.6	specific volume	m^3kg^{-1}
β	7.2, 9.9	coefficient of thermal volume expansion	K^{-1}

14.3

Symbol	Section	Quantity	SI unit
γ	7.3	ratio of specific heats	-
γ, γ_w	ch 12	unit weight of soil, water	Nm^{-3}
δ	8.4	isotope ratio variation	-
δ, δ^*	9.5	boundary layer thickness, displacement	m
Δ	8.4	isotope fractionation factor	-
	9.6	dimensionless diameter	-
ε, ε_0, ε_r	6.1, 11.1	permittivity, free space, relative	Fm^{-1}, Fm^{-1}, -
ε	9.3	emissivity	-
	9.4	eddy viscosity	Nsm^{-2}
	13.1	strain	-
ε_p	6.6	plastic straining range	-
ξ	3.6	critical damping ratio	-
θ	many	angle	rad
	8.6	potential temperature	°C
	9.5	boundary layer momentum thickness	m
$\dot{\theta}$, $\ddot{\theta}$	5.2	angular velocity, acceleration	$rads^{-1}$, $rads^{-2}$
λ	8.4	radioactive decay constant	s^{-1}
	9.3	wavelength	m
	9.9	mean free path	m
μ	5.2	coefficient of friction	-
	7.2, 7.3, ch 9	dynamic viscosity	Nsm^{-2}
μ, μ_0, μ_r	11.1	permeability, free space, relative	Hm^{-1}, Hm^{-1}, -
ν	7.1, 13.1	Poisson's ratio	-
	ch 9	kinematic viscosity	m^2s^{-1}
ρ	many	density	kgm^{-3}
ρ, ρ_0	11.1	resistivity	Ωm
σ	6.6, 12.4, 13.1	direct stress	Nm^{-2}
	7.2, 9.9	surface tension	Nm^{-1}
	9.6	Cavitation number	-
	2, 9.3	Stefan-Boltzmann constant	$Wm^{-2}K^{-4}$
σ'	8.6	density anomaly	kgm^{-3}
	12.4 - 12.6	effective stress	Nm^{-2}
σ_{th}	9.6	cavitation number (Thoma)	-
σ_y, σ_f	7.1	yield (proof), failure (ultimate) stress	Nm^{-2}
τ, τ_0	9.4, 9.5, 13.1	shear stress, surface (wall) shear stress	Nm^{-2}
ϕ	many	angle	rad
	5.2	angle of friction	
Φ	11.1	magnetic flux	Wb
ω, $\underline{\omega}$	ch 5, 9.4	angular velocity (vector)	$rad\ s^{-1}$
ω, ω_0	3.6, 11.1	circular frequency, natural	$rad\ s^{-1}$
ω_s	9.6	dimensionless specific speed	-

14.4

15. Index

Printed and bound by CPI Group (UK) Ltd, Croydon, CR0 4YY